Voyage of Discovery to Terra Australis

by Willem de Vlamingh in 1696–97

Dese Rivier 10 a: 12 Mijl=
Op Geweest En al Aan Brack
Water Bevonden:
A N D T
Vol
Boomen En
Eylandt ROTTENEST
Duytsche Mijlen is Voor Een Graadt
1 2 3 4 5 10 15 20
ANNO 169

Voyage of Discovery to Terra Australis

By Willem de Vlamingh in 1696–97

PHILLIP PLAYFORD

The National Library of Australia
Cataloguing-in-Publication entry:

Playford, Phillip E. (Phillip Elliott).
Voyage of discovery to Terra Australis, by Willem De Vlamingh in 1696-97.

Includes index.
ISBN 0 7307 1221 4.
ISBN 0 7307 2602 9 (pbk).

1. Vlamingh, Willem de, b. 1640 – Journeys – Western Australia.
2. Western Australia - Discovery and exploration. I. Title

919.41041

Edited by Nicole Blansjaar

Published by the Western Australian Museum
Francis Street, Perth, Western Australia 6000

Designed by Robyn Mundy, Mundy Design
Printed by Frank Daniels Pty Ltd

Frontispiece and cover: Willem de Vlamingh's three ships, the Nyptangh*, the* Weseltje*, and the* Geelvinck *(left to right), approaching Rottnest Island and the Swan River. From the map of the South Land charted by Vlamingh and drafted by Victor Victorszoon. Courtesy of the Algemeen Rijksarchief, The Hague.*

Contents

LIST OF FIGURES

Acknowledgments

I would like to express my appreciation to Cornelis de Heer and Christopher Clerk for their translation of Willem de Vlamingh's journal, which is included here as an appendix. This translation is the most recent of many ways in which Cornelis de Heer has generously assisted my historical research during the past 40 years.

Thanks are due to Lennie McCall of the Battye Library for supplying a copy of part of Madame Hélouis' transcription of the French translation of Vlamingh's journal, and to Helen Wade, Kathryn Dan, and Graeme Powell of the National Library of Australia for providing the rest of that transcription, together with comments on the origins of Madame Hélouis' work. I would also like to thank Maïté Etchechoury and Phillipe Henrat of the Archives Nationales de France, for arranging microfilming of the French journal and providing comments on it. Vincent Mollet of the Service Historique de la Marine, Vincennes, kindly facilitated the microfilming of a second contemporary copy of the journal.

The Algemeen Rijksarchief has assisted this project in a number of ways, especially by providing photographic copies of various maps. Femme Gaastra of Leiden University arranged for the Scheepvaartmuseum, Amsterdam, to provide a copy of Witsen's account of Vlamingh's voyage, and J. P. Sigmond of the Rijksmuseum gave information on the Dirk Hartog plate and arranged for x-ray images of it to be supplied. The Fisher Library, Sydney University, provided me with a copy of Valentyn's description of Vlamingh's voyage, together with a photograph of the engraving depicting the mouth of the Swan River. Volunteer library staff of the Royal Western Australian Historical Society are thanked for providing access to the Society's records relating to Dirk Hartog Island.

I wish to express appreciation to Arthur Hoffman, Murray Jones, Simon Urbini, and Peter Carroll of the Geological Survey of Western Australia, for their generous assistance in preparing the illustrations. My editor, Nicole Blansjarr, deserves thanks for her skilled editing of the manuscript, as do Robyn Mundy for her creative designing and Ann Ousey for facilitating the publication.

Several members of the Willem de Vlamingh Tricentennial Committee gave valuable support and encouragement, and among them I would especially like to express my appreciation to George Seddon, Peter Lynn, Graeme Henderson, and Brian Easton. I am also grateful to George Seddon, Reg Appleyard, and Graeme Henderson for their comments offered after reading the draft manuscript.

Members of the Dirk Hartog Island subcommittee deserve thanks for the time and effort they devoted to organizing the Dirk Hartog Island event as part of the Tricentennial celebrations, and in that respect I would especially like to thank Les Moss, Mike McCarthy, Tom Perrigo, and Jane Blake. Special appreciation is due to Alan Biggs, CEO of the Shire of Shark Bay, other staff of the Shire, and many residents of Denham who assisted in running the Dirk Hartog Island function. Geoff Wardle and his family are also thanked for their contributions to the function and for their hospitality towards my wife and myself during our preliminary visit to the island.

I also wish to record my indebtedness to Günter Schilder, whose comprehensive publications on Willem de Vlamingh provided much of the background information on which this study is based.

Finally I would like to thank my wife, Cynthia, for her love, encouragement, and tolerance, without which this could never have been written.

Prologue

AN AWAKING INTEREST

LIKE most Western Australians of my generation, I first heard about Willem de Vlamingh and his visit to Western Australia when I was at primary school. It was one of those memorable tales of Australian history that was told to us all. We learned how in 1696–97, this famous Dutch navigator had discovered and named the Swan River and Rottnest Island, and had found a pewter plate on Dirk Hartog Island, inscribed with a record of the landing there by Dirk Hartog in 1616. We also heard how Vlamingh had removed Hartog's plate and replaced it with one of his own, which was later found by French explorers and taken to Paris for safekeeping.

Near the end of my school years, the Vlamingh plate was given to the Australian people by the French Government and was placed on display in the Public Library, Perth. I remember being fascinated to see it there, alongside the original Vlamingh and Hamelin posts.

My early interest in Willem de Vlamingh and his voyage was reawakened in 1954, when as a young man I began a study of the *Zuytdorp*, one of the great ships of the Dutch East India Company, which had been wrecked south of Shark Bay in 1712. The story of Vlamingh's visit was especially relevant to that research, as it had taken place only fifteen years before the *Zuytdorp* was lost.

Soon after this I became further involved in the story when examining the geology of Dirk Hartog Island for West Australian Petroleum (WAPET). On my first visit to the island, in 1955, I climbed the cliff at Cape Inscription, seeking to follow the precise route that had been taken by the Dutch and French explorers so long before. At the cliff top, I could see the prominent cleft in the limestone where the famous plates and posts had been erected in 1616 and 1697, and felt deeply moved by the sense of history at this lonely and imposing place.

In subsequent years, I often reflected on my visit to Cape Inscription, and in 1991, I realised that in five years' time it would be 300 years since Willem de Vlamingh's voyage of discovery along our coast. As chairman of the Rottnest Island Environment and Research Committee, I proposed that celebrations be held to mark this significant occasion. My proposal was adopted by the committee and the Rottnest Island Authority, and in due course a series of commemmorative events was held, during the period December 1996 to February 1997.

The committee set up to organise the tricentennial celebrations believed that a new popular account of Willem de Vlamingh's visit to Western Australia was warranted, and as I was already gathering information for a public lecture on Vlamingh, it was eventually concluded that I should also put together a book. This was originally supposed to be published in 1997, but its completion was delayed because of my discovery in February 1997 of an important document that had been overlooked by other recent scholars—a handwritten French translation of Vlamingh's personal journal, the original Dutch version of which has long disappeared. This journal provides invaluable insights into the man and his voyage, and its rediscovery led to extensive revision of the manuscript, thereby delaying publication by more than a year.

The manuscript was finally completed in 1998, 301 years after the end of Willem de Vlamingh's epic voyage. I hope that readers will excuse my cliché of 'better late than never', and that the story will prove as interesting to you in reading as it has been to me in writing.

In search of a lost ship

ORIGINS OF THE EXPEDITION

WILLEM de Vlamingh's voyage of exploration to Terra Australis was proposed primarily to investigate the loss of the Dutch East India Company's ship *Ridderschap van Holland*, which had disappeared after leaving the Cape of Good Hope on 5 February 1694, bound for Batavia (Jakarta). The vessel had on board Sir James Couper, an eminent official of the company who was due to take up an appointment on the Governing Council in Batavia. Early speculation on the loss of the ship focused on the likelihood that it had been wrecked on the coast of Western Australia, then known variously as Terra Australis, the South Land, the Land of the Eendracht, and New Holland.[1] This little-known coast had previously claimed two of the company's vessels, the *Batavia* in 1629 and the *Vergulde Draeck* ('Gilt Dragon') in 1656.

In an endeavour to solve the mystery of the missing ship, the directors of the Dutch East India Company ('de Verenigde Oost Indische Compagnie', or VOC) decided in November 1695 that a search expedition would be sent to look for wreckage and rescue any survivors.[2] The expedition would also seek castaways from the *Vergulde Draeck*, who had been left behind on the coast of Terra Australis in 1656, even though it seemed unlikely that any could still be alive after forty years. On the way to the southern continent, the expedition vessels were to explore the islands of Tristan da Cunha in the South Atlantic and Saint Paul and Amsterdam in the southern Indian Ocean. They would then go on to carefully explore the west coast of the continent, looking for any wreckage and survivors, and determining whether anything of economic value could be obtained there. Indeed, exploration of that little-known land seems to have become the expedition's foremost objective in the mind of its principal promoter, Nicolaes Witsen, a director of the VOC and burgomaster (mayor) of Amsterdam, who was also well known as a scholar, scientist, and author.

Willem de Vlamingh was selected as commodore of the expedition and skipper of the lead vessel, the *Geelvinck*, a newly constructed frigate with a crew of 134 men. The name '*Geelvinck*' means 'Yellow Finch', and such a bird was painted on the stern of the vessel. However, it was also the surname of a director of the VOC who assisted Witsen in drawing up the sailing instructions for the expedition, and the ship was presumably named in his honour. The other two vessels in Vlamingh's flotilla were the hooker *Nyptangh* ('Nipper' or 'Pincers'), with a crew of fifty men, skippered by Gerrit Colaert, and the galliot *Weseltje* ('Little Weasel'), with a crew of fourteen men, skippered by Laurens Teunisz Zeeman. Vlamingh's son, Cornelis de Vlamingh, aged eighteen, began the voyage as third mate of the *Geelvinck*, and was promoted to command of the *Weseltje* after the death of Zeeman at the Cape of Good Hope.[3]

Willem de Vlamingh was born in 1640 on the island of Vlieland, one of the West Friesland Islands (see Figure 1). He was christened in the Dutch Reformed Church in the town of Oost Vlieland, his full name being Willem Hesselszoon de Vlamingh van Oost Vlieland (or, in English, 'William son of Hessel the Fleming of East Vlieland'). His family presumably adopted the name Vlamingh ('Fleming' in English) because they originally came from Flanders in the Spanish Netherlands. They may have fled to the Dutch Republic during the late sixteenth century to escape the reign of terror being waged against Protestants by the Spanish.[4]

Figure 1. *Map of the Dutch Republic and the Spanish Netherlands in the early seventeenth century. Willem de Vlamingh (William the Fleming) was born on the island of Vlieland; his family presumably came originally from Flanders in the Spanish Netherlands.*

Willem de Vlamingh had previously made a name for himself as a whaler and explorer of the arctic seas north of Russia. He had joined the Dutch East India Company in 1688 and completed two trading voyages to the East Indies before being selected to lead the expedition to Terra Australis. He was fifty-six years old when the expedition began.[5]

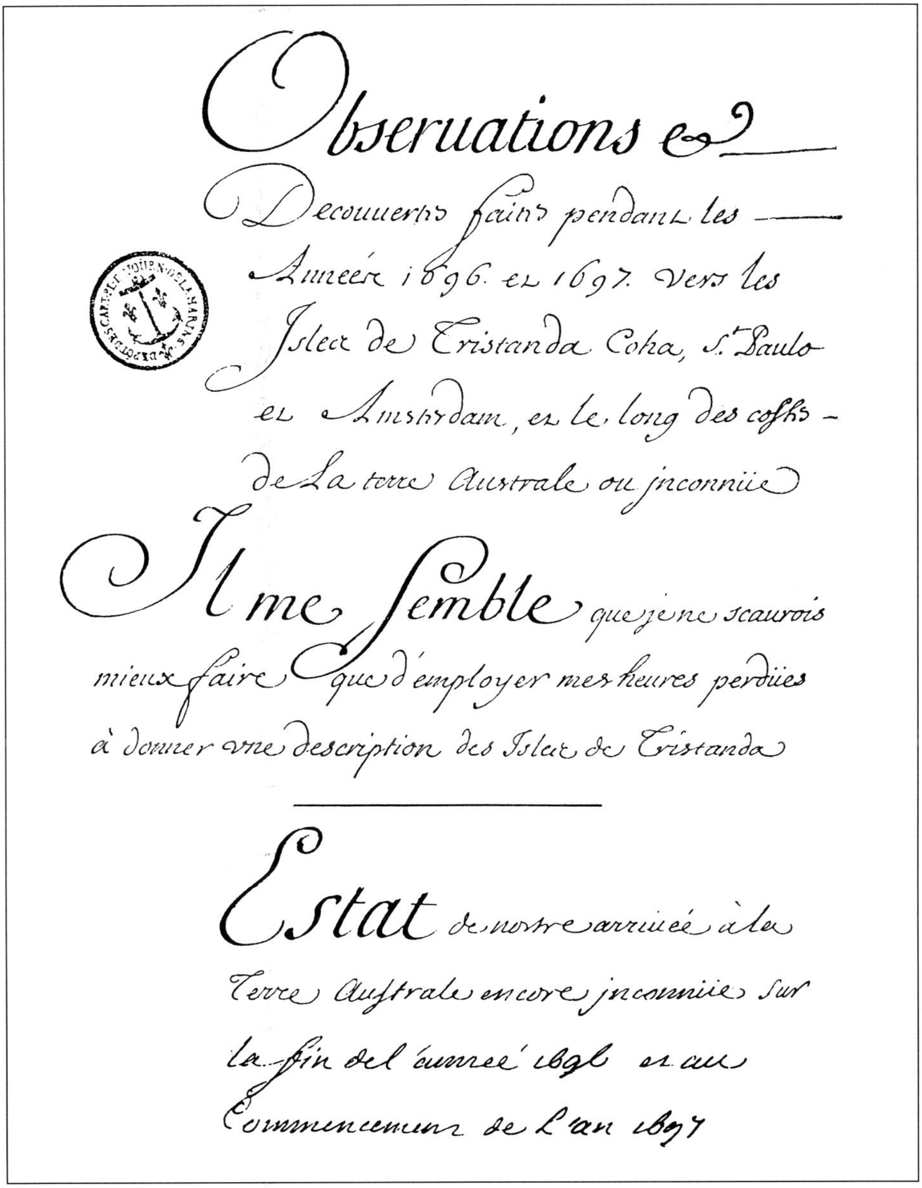
Observations &
Decouvertes faites pendant les
Années 1696 et 1697 vers les
Isles de Tristanda Coha, S.t Paulo
et Amsterdam, et le long des costes
de la terre Australe ou inconnüe

Il me semble que je ne scaurois
mieux faire que d'employer mes heures perdües
à donner une description des Isles de Tristanda

Estat de notre arrivée à la
Terre Australe encore inconnüe sur
la fin de l'année 1696 et au
Commencement de l'an 1697

Figure 2. *Extracts from the French translation of Willem de Vlamingh's journal, showing the title and beginning of the journal (top and middle paragraphs) and the title of the section dealing with Terra Australis (bottom paragraph).*

Courtesy of the Archives Nationales de France, Paris

The most recent published accounts of the voyage are a series of books and maps published by Schilder in 1976, 1978, 1984, and 1985, and an earlier book by Robert published in 1972. These were based primarily on the logs of the *Geelvinck* and the *Nyptangh*, and did not take account of the French translation of Vlamingh's personal journal, which is thought to have been translated from the Dutch original (now lost) during the early eighteenth century. It is entitled (in English translation): 'Observations and discoveries made during the years 1696 and 1697 about the islands of Tristan da Cunha, Saint Paul, and Amsterdam, and along the coast of the Terra Australis or Incognita'.

J.S. Battye of the Public Library, Perth, acquired a transcript of part of this French manuscript in 1928 from the Parliamentary Library, Canberra, which had in turn obtained it in 1912 from Madame Robert Hélouis in Paris.[6] She had copied it by hand from the original French manuscript, then held in the Depôt de la Marine Service Hydrographique. The original manuscript is now stored in the Section Ancienne of the Archives Nationales de France in Paris, and another contemporary copy is held in the Marine Nationale, Service Historique, Vincennes.[7]

The French translation of Vlamingh's journal was first mentioned by Stapel in 1937, and an English translation of part of the text relating to Rottnest Island was published by Somerville in 1948.[8] Surprisingly, however, the journal was not discussed by subsequent scholars who have studied Vlamingh and his voyage.[9]

The Archives Nationales manuscript was translated into English for the present study by Cornelis de Heer and Christopher Clerk, and is included here as an appendix.[10]

It is interesting to note that Vlamingh, in his journal, preferred to use the name 'Terra Australis' ('*Terre Australe*' in French) when referring to the Australian continent, occasionally also using the name 'South Land' but never the names 'Land of the Eendracht' or 'New Holland'. The usage 'Terra Australis' dated from the time before the discovery of the west coast of the continent by Dirk Hartog in 1616.[11] Until then the mythical southern continent was generally known (in Latin) as 'Terra Australis Incognita', the Unknown

South Land. Clearly, as far as Vlamingh was concerned, the continent remained essentially unknown prior to his visit.

The name 'Terra Australis' would again be revived, 117 years later, by Matthew Flinders, who believed that it was more appropriate as a name for the continent than either 'New Holland' or 'New South Wales'.[12] However, Flinders, among others, also advocated the shortened name 'Australia', and this soon came into general use, after being used officially for the first time in 1817.[13]

The substance of Vlamingh's journal is similar to that of the *Geelvinck* log, but the two accounts differ in some significant ways. Schilder points out that Vlamingh's *Geelvinck* log is 'couched in the language of a sober sailor' and that Vlamingh 'shuns elevated diction and limits himself to simple statements of fact'.[14] By contrast, in his personal journal, Vlamingh freely expresses his inner thoughts and describes his own impressions of the strange new land of Terra Australis, often expressing his views in quite poetic terms. Consequently, his journal is of major historical importance.

The personal content of Vlamingh's journal leaves no doubt that it was written by the man himself, and Schilder has shown that the *Nyptangh* log was written by the ship's uppersurgeon, Mandrop Torst. However, there is some doubt regarding the authorship of parts of the *Geelvinck* log. Certainly, most of the log is worded as though it was prepared by Vlamingh, but there are some differences between equivalent parts of the log and Vlamingh's personal journal which imply that they were not written by the same person. I believe that some sections of the log were probably written by the ship's bookkeeper (or 'assistant'), Joannes Bremer, on behalf of the skipper. Perhaps other parts were written by the uppersteersman, Michiel Bloem.

Vlamingh may have drafted the manuscript of his journal soon after arriving in Batavia, using diary notes prepared during the voyage. He probably intended publishing it, but died before that could be achieved. His journal begins as follows:

> It seems to me that I cannot do better than to use my spare time to provide a description of the islands of Tristan da Cunha and their surroundings

and also the islands of St Paul and Amsterdam, and the discoveries we made about the Terra Australis or the South Land, which, up to the present, have neither been explored properly nor described. This work will, perhaps, be of some use to our heirs, and, although I cannot make it as polished in expression as I would wish, the reader will have the goodness to believe that I have nevertheless done my best for his benefit and satisfaction.

The scourge of scurvy

The Netherlands to the Cape

THE three ships commanded by Willem de Vlamingh left the island of Texel in the Netherlands on 3 May 1696, sailing 'strictly incognito' as part of a convoy of 130 vessels escorted by two warships. The Dutch Republic was at war with France at that time, and there would have been a danger of interception by French warships and privateers if the shorter route to the Atlantic through the English Channel had been followed. Consequently, they sailed via the safer 'back way', around the coast of Scotland, despite the additional time and distance involved (see Figure 3).

As instructed, the expedition touched briefly at the uninhabited islands of Tristan da Cunha in the South Atlantic, before proceeding to the Cape of Good Hope, which was reached by the *Nyptangh* on 3 September, the *Geelvinck* on 7 September, and the *Weseltje* on 8 September. By that time, many of the people on board the three ships had become seriously ill with scurvy, and several had died—the *Geelvinck* alone had seven dead and seventeen were hospitalised at the Cape.[15]

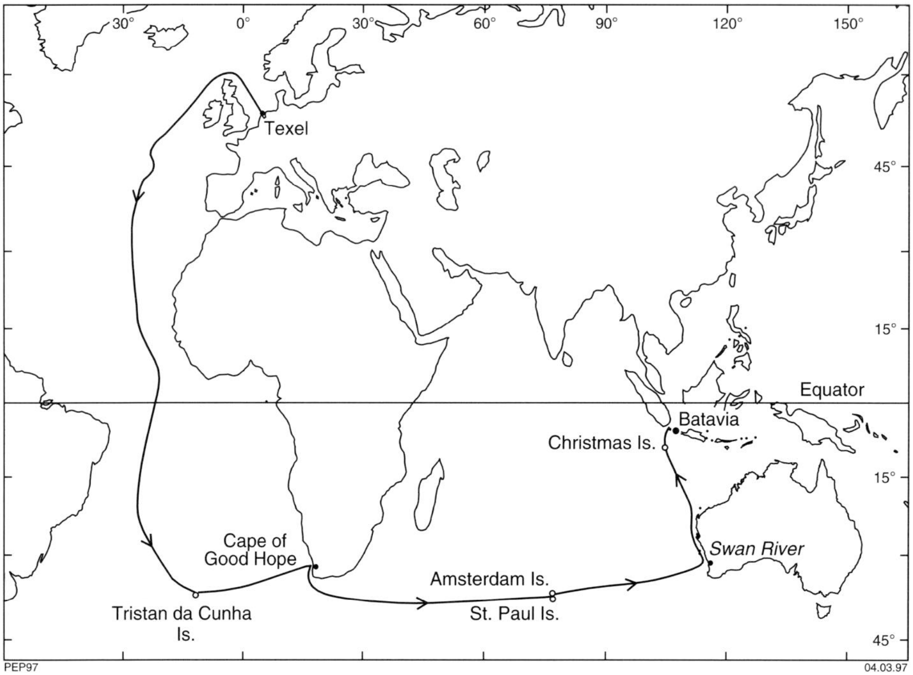

Figure 3. *Route followed by the three ships commanded by Willem de Vlamingh on the voyage from the Netherlands to Batavia in 1696–97.*

Scurvy was the scourge of seafarers at that time, and many sailors succumbed to this life-threatening disease on the long voyage from the Netherlands to the Cape of Good Hope. Scurvy resulted from the lack of fresh fruit and vegetables in the shipboard diet, leading to serious vitamin-C deficiencies. The uppersurgeon of the *Nyptangh* wrote (in the surgeon's journal) that by the time the ship reached Tristan da Cunha, more than three months after leaving the Netherlands:

> The people began to complain of fatigue and stiffness all over their bodies; it being the beginning of scurvy ... they fall into burning fevers, a pale and leaden colour, short and stinking breath, a constriction over the heart, little appetite, blue swollen rotten gums, blue and green spots over the

> whole body, and others like fleabites on the shins, rigidity through all the limbs, headaches, etc.[16]

The uppersurgeon of the *Geelvinck* described the symptoms of scurvy in his surgeon's journal as follows:

> Rigidity in the limbs and especially in the juncture of the jaws, with purple spots and a bluish rash, a constriction above the heart, short of breath, a pale complexion, and fatigue of the body, all this together with a sound heart, but difficulty in biting, the mouth being in a poor condition through spongy and rotten gums and loose teeth. I had their legs bathed every morning ... and internally administered twice a day these spirits, to be taken in brandy fortified with horseradish. Their nourishment was beer and bread or a rusk with butter or cheese, for those who could bite it, and a glass of sack, and for the rest whatever they liked and could be supplied on board. With this I sustained them until they arrived at the Cape of Good Hope.

The health problems on Vlamingh's ships, appalling as they seem today, were not unusual at that time, so when the ships reached the Cape, the official report stated that the three vessels had arrived 'without dead or sick people worth mentioning'. Indeed, people were regarded as expendable: on average only one person in three who left the Netherlands for the Indies ever returned to their homeland.[17]

The visit to the Cape must have been a godsend to the crew, as most would have been cured of scurvy during their stay by a diet that included fresh fruit and vegetables, for even in serious cases, recovery could be achieved in only two or three weeks.[18] However, from the VOC's viewpoint, the ships remained at the Cape too long — seven weeks, about double the time usually allowed — and Vlamingh would later be criticised by Witsen for wasting time there 'feasting and merrymaking'.[19] It is noticeable that nothing is said about the visit to the Cape in Vlamingh's own journal — perhaps he thought that it was better to remain silent on this matter.

While at the Cape, Vlamingh selected three 'black convicts' from the East Indies (described in the ships' logs as 'Indians') to join the expedition.[20] These men were proficient in Dutch, Portuguese, Malay, and many Indonesian dialects, and it was hoped that they would be able to communicate with the people of Terra Australis. Of course, the Dutch were unaware that the languages spoken by the Aborigines bore no resemblance to those of neighbouring countries.

The roaring forties

Across the Indian Ocean

THE three ships left the Cape about one month later than scheduled, on 27 October 1696, and sailed east with the prevailing westerly winds, following the route across the Indian Ocean that had been pioneered eighty-five years before by Commodore Hendrik Brouwer. The strong westerly winds, known as the 'roaring forties', carried the ships rapidly to the islands of Saint Paul and Amsterdam in the southern Indian Ocean. Crew members explored the two islands, but found no wreckage or signs of human activity.

The ships then sailed on towards Terra Australis, seeking to accomplish the most important part of their mission: exploration of that little-known continent. Vlamingh had been instructed not only to search for people or wreckage from the *Ridderschap van Holland*, the *Vergulde Draeck*, and any other lost ship, but also to explore thoroughly the coast and as much as possible of the interior. He was to ensure that guard was maintained against the 'very savage, barbaric, and cruel people' who were said to inhabit the southern continent. On the other hand, Nicolaes Witsen had asked Vlamingh to 'catch a South Lander and bring him hither'.[21] The intent was no doubt to teach the Dutch language to the captive, so that he could advise the VOC regarding any commodities in his country that might be of economic value.

It probably never occurred to Witsen that such an abduction would have been 'barbaric and cruel'; he, like most other Europeans at that time, applied such terms to 'uncivilised' native people, not to 'civilised' whites. The VOC had from time to time issued formal instructions forbidding the kidnapping of native people, but these were disregarded with impunity by many mariners, no doubt with the unwritten approval of the company. For example, personnel involved in the kidnapping of Aborigines from Cape York Peninsula in 1623 and 1756 would never have described those seizures in their ships' logs unless they believed that their actions would receive tacit approval.[22] There was no mention in the formal instructions issued to Vlamingh of any intention to kidnap Aborigines, but it is clear from Witsen's statement and Vlamingh's actions that members of the expedition would endeavour to 'catch' and 'lay aboard' at least one of those people. However, the Aborigines avoided direct contact with the explorers by hiding or fleeing when they approached, thereby avoiding any incidents.

Paradise on earth

ROTTNEST ISLAND

ON 28 December, 'green stuff' was seen floating by (*Nyptangh* log), and in his journal, Vlamingh reported seeing a 'quite extraordinary bird', which he identified as a gull, having 'a black head and wings, the rest of the body being completely white'. From this he judged that they were not far from the coast.

Following this sighting, Vlamingh directed that the lead (sounding line) be cast often, in accordance with company regulations when approaching the southern continent. This was done in the expectation that shallow water would be found near the coast, thus providing warning of its proximity. The relevant regulation stated that:

> The Land of the Eendracht has south of 27 degrees many perilous shallows and sharp grounds, therefore cautious action is needed and timely watch must be kept, and at night and in darkness it is urgent to use the lead; bottom is found at 100, 80, and 70 fathoms for 7, 6, and 5 miles [50–35 kilometres] off the coast.[23]

Bottom was eventually reached sounded at 65 fathoms (110 metres), when the *Geelvinck* would have been about 45 kilometres off the mainland coast. Vlamingh stated in his journal that 'this gave us extraordinary joy, the more so as the great God had favoured us with unparalleled care, and, to glorify him, we allowed ourselves to drift for a period of four hours'.

Next day, Saturday, 29 December, they saw floating tree branches and stumps, and at one o'clock in the afternoon land was sighted for the first time (see Figure 4). The *Geelvinck* log recorded that they 'saw the Fog Island (*Misteiland*) dead ahead'. This island, now known as Rottnest Island, is referred to by four names in the expedition records: 'Fog Island' (*Geelvinck* log), the 'Isle of Girls' (Vlamingh journal), 'Rats' Island' (*Nyptangh* log) and 'Rottenest Island' (map and sketch by Victor Victorszoon — see Figures 5 and 7).

The reason why it was referred to in the *Geelvinck* log as 'Fog Island' is unclear. The island was sighted near the middle of a summer's day, when fog would be most unlikely. Moreover, there is no mention of fog in the ships' logs or Vlamingh's journal. Perhaps the island loomed into view through a humid haze of smoke blown from the mainland, where many bushfires, lit by the Aborigines, were then burning.

Another name for the island, 'the Isle of Girls', is given in the French translation of Vlamingh's journal (as '*l'Isle des filles*'). The origin of this name is also obscure — certainly there were no girls on the island at the time of the Dutch visit! A possible explanation is that the name has resulted from a translation error. The original Dutch version of the journal is lost, so we cannot be sure, but perhaps Vlamingh's handwriting was poor or the manuscript was damaged. The French translator may have misread the word '*Misteiland*' as '*Meisjeseiland*' (Girls Island). Another hypothesis, suggested by Stapel, is that the name was misread as '*Misseiland*'. As that is not a Dutch word, Stapel thought that the translator might have concluded that Vlamingh was using the English word 'miss', meaning 'girl'.[24] However, neither of these explanations seems particularly convincing.

Apart from its use in the French manuscript translation of Vlamingh's diary, we know that the name '*I. des filles*' (Isle of Girls) was used elsewhere in France during the eighteenth century. It was the name given to the island on a

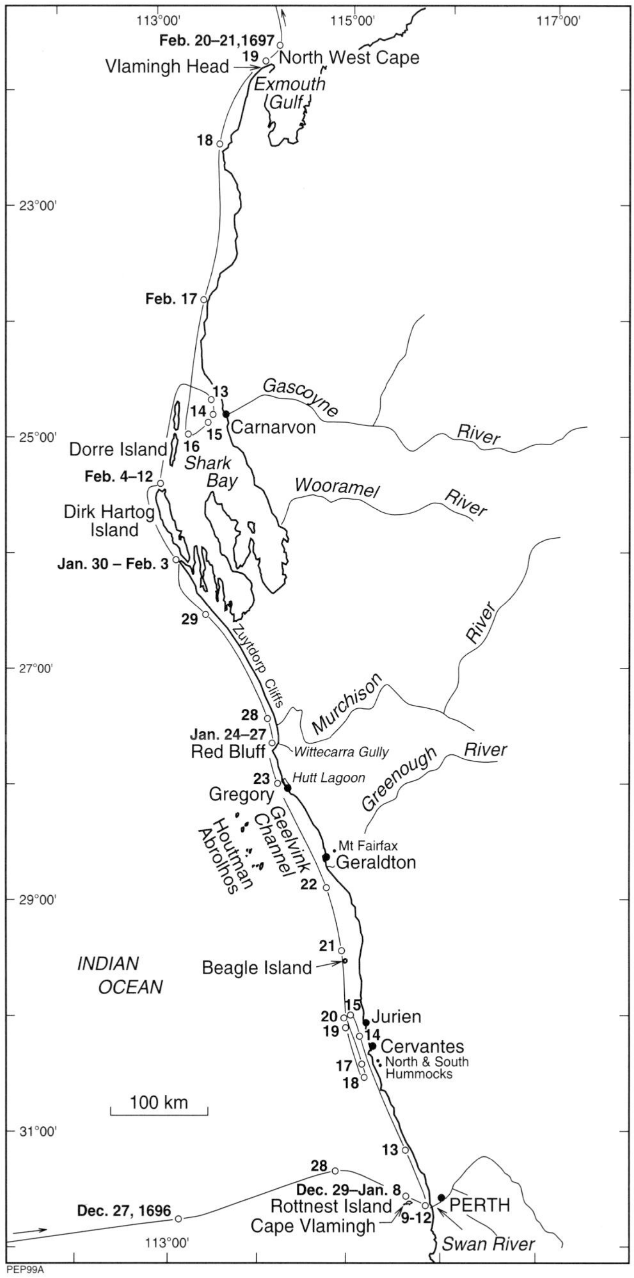

Figure 4. Route followed by the three ships commanded by Willem de Vlamingh along the Western Australian coast during the period December 1696 to February 1697.

French 'reduced chart of Terra Australis' produced by Sr. Bellin, a marine engineer of the Royal Society of London, in 1753 and published by Prévost d'Exiles in 1758 (see Figure 8).[25] Schilder concluded, enigmatically, that this map 'is chiefly interesting as a demonstration of the French penchant for hypotheses'.[26]

The references to 'Fog Island' in the *Geelvinck* log and 'Isle of Girls' in Vlamingh's journal are made without explanation, as though each name was already known. It is possible that one or both names had previously been given to the island following its discovery in 1658 by Samuel Volkersen of the ship *Waeckende Boey*, who was at that time searching for survivors of the *Vergulde Draeck*. Volkersen said in his journal that he had been unwilling to give a name to 'this large island', 'leaving that matter to the Hon. Governor-General's pleasure'.[27] Perhaps the governor-general and someone else in the VOC independently gave the island the names 'Fog Island' and 'Isle of Girls', for reasons that are now unknown.

In the *Nyptangh* log the island is referred to as 'Rats' Island', no doubt because of the many 'bush rats' (marsupial quokkas) that expedition members found there. The fourth name, '*'t Eylandt 't Rottenest*', meaning literally 'the Island the Rats' Nest', clearly had the same derivation. That name was shown on the magnificent map of the *Zuyd Landt* (South Land), drawn by Victor Victorszoon, the *Geelvinck*'s artist, draftsman, and 'consoler of the sick' (see Figures 5 and 6). He also wrote the name on his watercolour sketch of the north shore of the island (see Figure 7). Victorszoon's name has prevailed to the present day, in a slightly modified form, as 'Rottnest Island'.

Victorszoon made watercolour sketches of fifteen localities on the coast of the South Land, the first being Rottnest Island. These sketches and Victorszoon's map of the South Land were all supposed to have been lodged in the VOC collection, and although that was done with the map, the sketches were retained by Claas Bichon, the commodore of the ship that carried them from Batavia to the Netherlands in 1697. The sketches were carefully preserved by the Bichon family for 273 years, until they were finally donated in 1970 to the Prins Hendrik Museum in Rotterdam. At that time, the Bichon family apparently knew little about the origin and identity of the sketches. They were identified by Günter Schilder soon after being lodged at the museum.[28]

Victorszoon introduced other geographic names through inscriptions on his map, and his names Swan River (*Swane Rivier*), Dirk Hartog Island (*Dirck Hartoogs Eylandt*), Red Bluff (*Roode Houk*), Steep Point (*Steyle Houck*), and Dorre Island (*Dor Eylandt*) are still in use. None of these names are mentioned in the ships' logs or Vlamingh's journal, but the Swan River was referred to in a VOC document written from Batavia in November 1697.[29]

The *Geelvinck* anchored off the north coast of Rottnest Island, about 2 kilometres north of The Basin, on 29 December 1696. The hooker *Nyptangh* and the galliot *Weseltje* may have anchored nearby, although according to the *Nyptangh* log, that vessel was moored on the south side of the island. Previous accounts of Vlamingh's voyage have suggested that the *Geelvinck* also anchored there, in either Porpoise Bay[30] or Thomson Bay.[31] However, my examination of the watercolour sketch of the north side of the island, painted by Victor Victorszoon on board the ship lying at anchor,[32] indicated that the mooring point was at neither of those places, but was located about 2 kilometres north of The Basin on the north side of the island (see Figures 5, 6, 7, and 9).

To confirm this, Peter Lynn took me to the predicted locality in his launch on 17 November 1995, accompanied by George Seddon, Thom Dercksen, and Mike Flood. There could be no doubt that we were at the *Geelvinck*'s anchoring site, as from that spot we could precisely line up several prominent features shown on Victorszoon's sketch, including Phillip Rock, Phillip Point, Thomson Bay, Duck Rock, and Bathurst Point (see Figures 7 and 9).[33] We also found that the water depth was the same as that recorded by Victorszoon (10 Dutch fathoms, or 17 metres).

Looking towards the island from Vlamingh's anchoring point we deduced that crew of the *Geelvinck*, when setting out for the island by boat, would probably have headed towards the conspicuous white sand dunes that could be seen at the eastern end of Longreach Bay. We motored towards those dunes in the launch, and soon found that a continuous offshore reef bars the bay there, making it impossible for a boat to land. However, looking a short distance to the east, a channel could clearly be seen passing through the reef to the

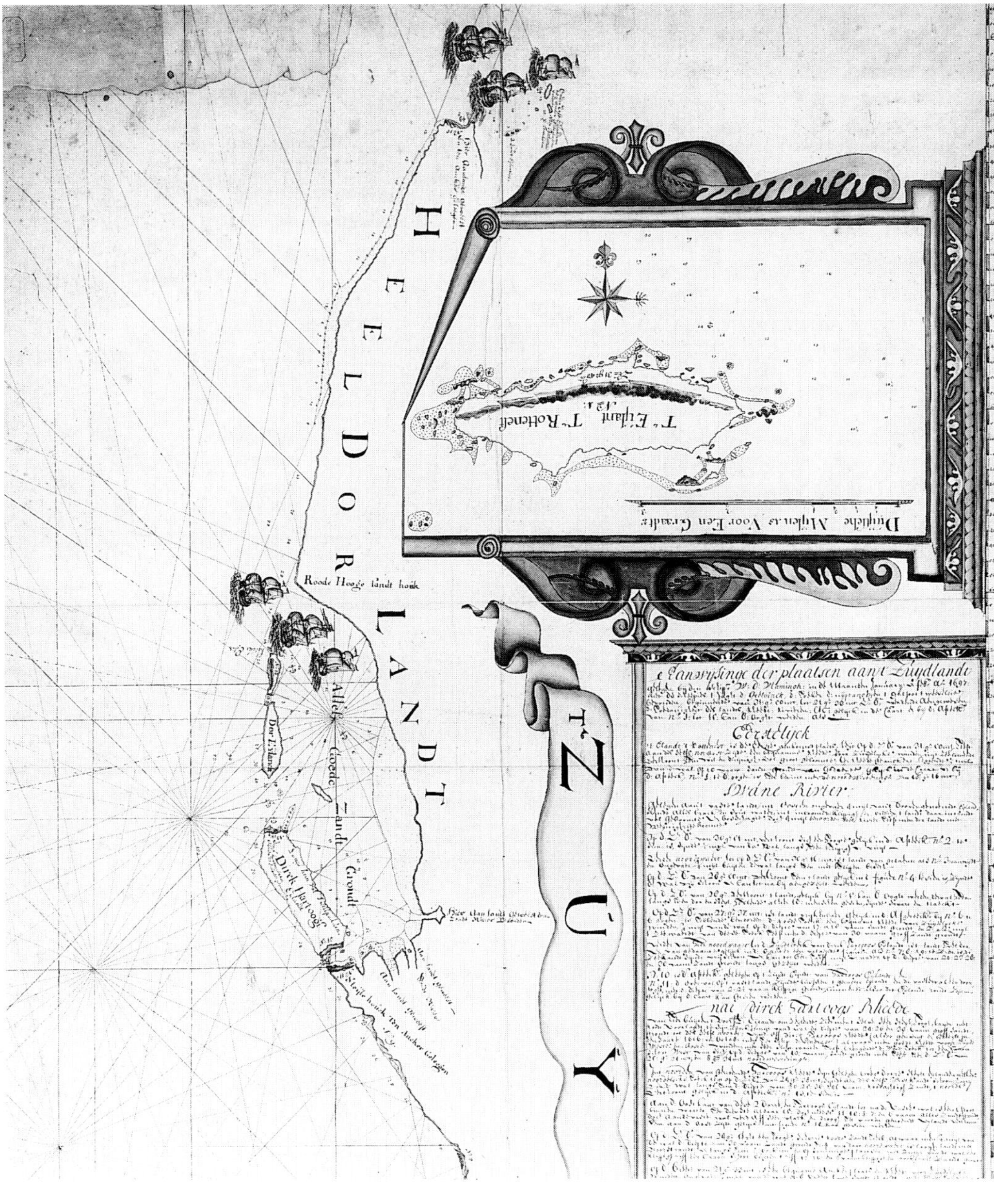

***Figure 5.** Northern part of the map of the South Land and detailed map of Rottnest Island, charted by Willem de Vlamingh and drafted by Victor Victorszoon.*

Courtesy of the Algemeen Rijksarchief, The Hague

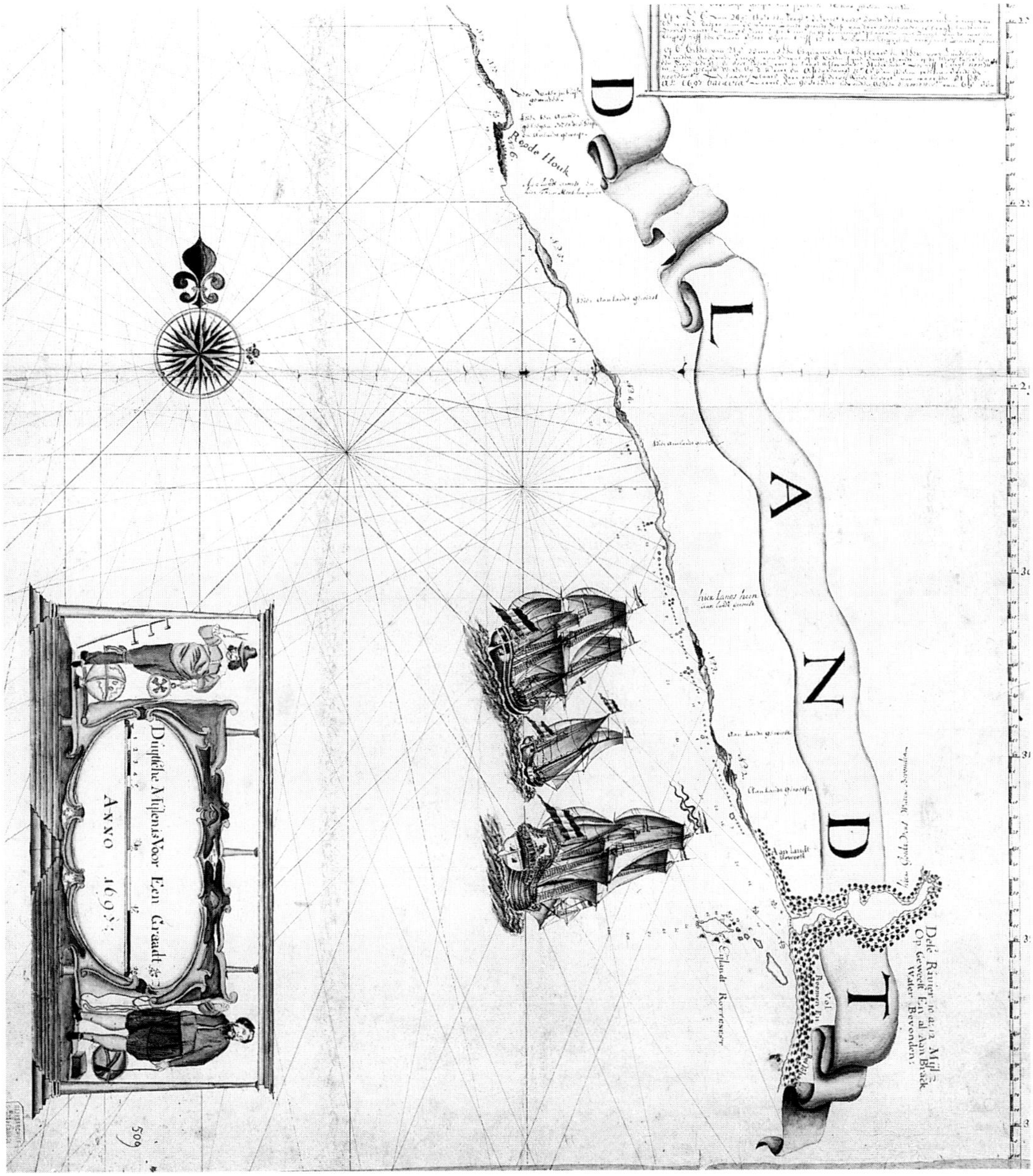

Figure 6. *Southern part of the map of the South Land, charted by Willem de Vlamingh and drafted by Victor Victorszoon.*

Courtesy of the Algemeen Rijksarchief, The Hague

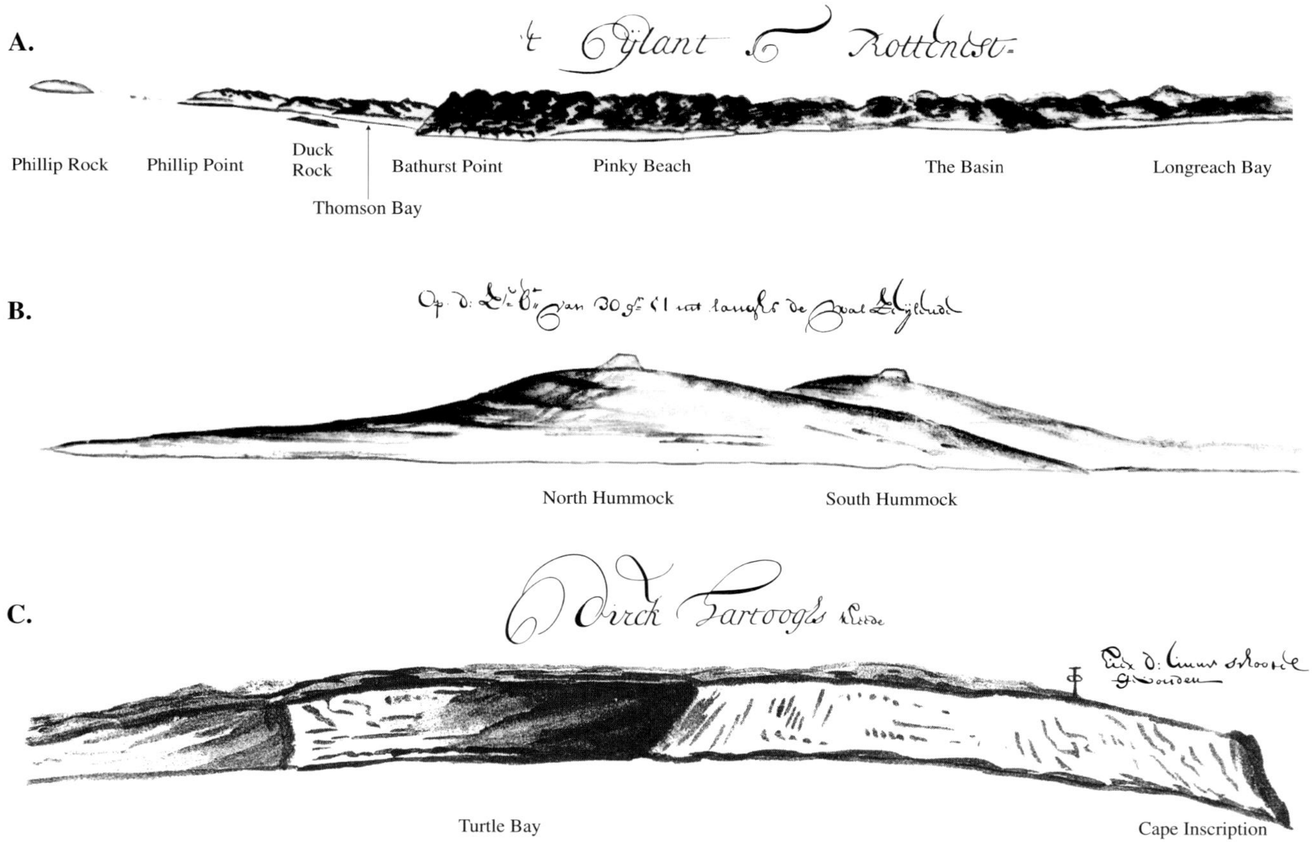

Figure 7. *Copies, with modifications, of parts of sketches by Victor Victorszoon: (a) the north side of Rottnest Island as viewed from the anchoring point of the Geelvinck; (b) the coastline south of Cervantes, showing North and South Hummocks; (c) the north end of Dirk Hartog Island, showing the place where Dirk Hartog's plate and post were found.*

After reproductions by Schilder (1984) of the original watercolour sketches held in the Prins Hendrik Museum, Rotterdam beach at The Basin. We felt confident that the Dutchmen would have decided that this was the best place to go ashore.[34]

Victorszoon's map of the South Land includes, as an inset, a detailed map of Rottnest Island. On this map the *Geelvinck*'s mooring point on the north side of the island is marked with an anchor symbol (see Figure 5). Victorszoon was a skilled artist and draftsman, but he was not a professional cartographer, and he did not adhere to the convention of orienting all captions right-way up in relation to north. For some unknown reason he drew the captions upside down on the inset map of Rottnest Island, even though the north pointer was conventionally directed upward and the water depths were written with normal orientations. Subsequent cartographers apparently did not notice that when

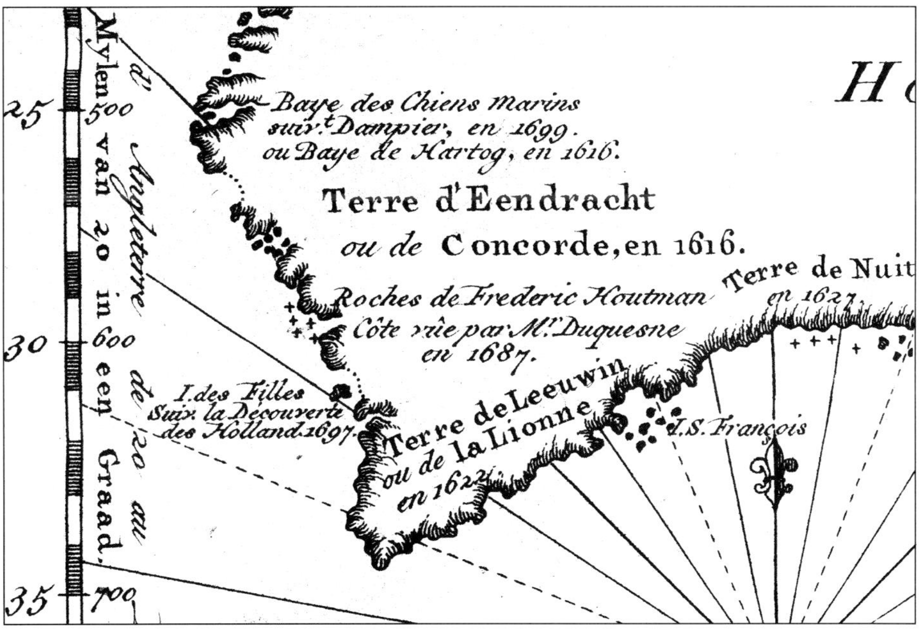

Figure 8. *Part of a map of Terra Australis, drafted by Bellin in 1753 and published in 1758. It shows Rottnest Island as* I. des Filles *('Isle of Girls'), the same name as that used in the French translation of Vlamingh's journal.*

Courtesy of the State Reference Library, Perth

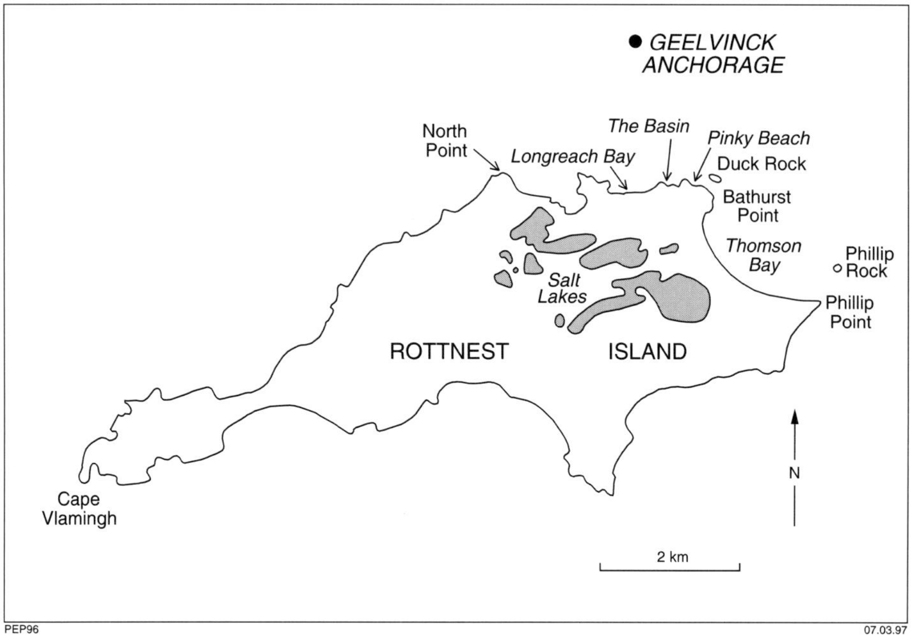

Figure 9. *Map of Rottnest Island, showing the place where the* Geelvinck *anchored on 29 December 1696.*

this inset map is turned around to read the name of the island, the north pointer is directed downward.

In the late seventeenth or early eighteenth century, Isaak de Graaf and Gerard van Keulen copied Victorszoon's map of the South Land. However, they drew his inset map of Rottnest Island as viewed with the name right-way up, without realising that north should then have been directed downward. Consequently, the island was inadvertently oriented upside down on their maps. Their mistake was repeated on the first published map of the island, by Johannes van Keulen (son of Gerard) in 1753 (see Figure 10).[35] Gerard and Johannes van Keulen were aware that the *Geelvinck* (and probably the other two ships) had been anchored off the north-east side of the island, so they added three anchor symbols in what they thought was that position, without realising that they were in the wrong place because their map was drawn upside

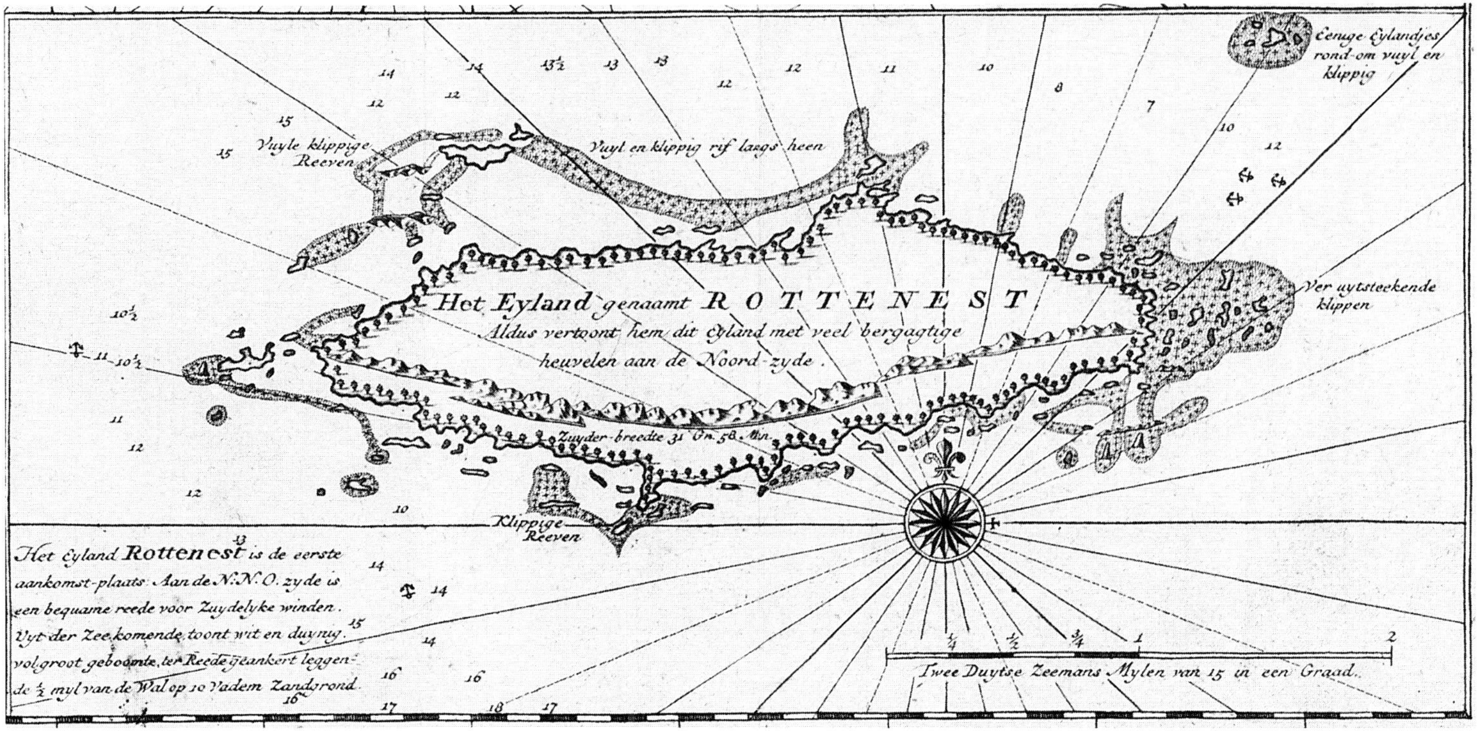

Figure 10. *Map of Rottnest Island, from the map of the Land of the Eendracht or New Holland, published in 1753 by Johannes van Keulen. The map of Rottnest Island is taken directly from the 1697 map of Victor Victorszoon, except that Johannes van Keulen has inadvertently drawn the island upside down.*

Courtesy of the National Library, Canberra

down. It is surprising that these significant errors have not been recognised by other authors who have made detailed studies of Vlamingh's voyage.

Vlamingh reported in his journal that he and his crew had intended going ashore after anchoring off the island, in order to 'amuse ourselves with exploring the island and admiring the fine trees which we saw in the distance, [but] the darkness of night robbed us of this pleasure'.

Next morning, 30 December, the longboats from the *Geelvinck* and *Nyptangh* sailed and rowed together around the island, searching for signs of castaways or wreckage. They returned without finding anything of note.

On the same day, the bookkeeper, uppersteersman, commander of the soldiers, and twelve others from the *Geelvinck* went ashore on the island in a pinnace, probably landing at The Basin. They returned to report having seen many 'rats nearly as big as cats', several of which they killed. Later, Witsen reported that the animals had 'a pouch below their throats into which one could put one's hand, without being able to understand to what end nature had created the animal like this'.[36] Of course, the Dutch seamen's 'rats' were the island's renowned quokkas (small marsupials), which seem to have been as common in 1696 as they are today. The *Nyptangh* journal recorded that the animals' droppings were 'found in abundance all over the island'.

More parties were sent ashore during the next several days to cut firewood for the galley and the *Geelvinck*'s water-distillation plant. One party found a piece of wood from a Dutch ship on the beach, with nails still sticking in it, and another saw a fragment of wood panelling. This wreckage probably came from the *Vergulde Draeck*, which had been lost forty years before on a reef 100 kilometres north of the island.[37] Another party reported finding several salt lakes in the centre of the island, describing them as 'basins with exceedingly salt water'. However, they said that 'water fit to drink' could be obtained by digging only seven paces from the basins.[38]

Willem de Vlamingh himself went ashore with the bookkeeper, Joannes Bremer, on 31 December. Landing without difficulty, they set out to explore the island, and after walking for about an hour came upon a thick and attractive grove of trees about half a Dutch mile (3.7 kilometres) long.[39] In his journal, Vlamingh observed that:

> Its trees were arranged in such a way that one would think they had been planted in a straight line by a professional, to make it into a park. From these trees we could smell a very pleasant fragrance, which was spread all over the island.

Vlamingh said that he would have liked to press on further, but as he had already become tired from the short walk, he asked the bookkeeper to continue on his own. Vlamingh turned back to rejoin other crew members, who were cutting wood, and he subsequently made the following enthusiastic comments about the island in his journal:

> On my way back I had great pleasure in admiring this island, which is very attractive, and where it seems to me that nature has denied nothing to make it pleasurable beyond all islands I have ever seen, being very well provided for man's well-being, with timber, stone, and lime for building him houses, only lacking ploughmen to till these fine plains. There is plentiful salt, and the coast is full of fish. Birds make themselves heard with pleasant song in these scented groves.
>
> So I believe that of so many people who seek to make themselves happy, there are many who would scorn the fortunes of our country for the choice of this one here, which would seem a paradise on earth.

Vlamingh was clearly enchanted by the island. However, Mandrop Torst, writing in the *Nyptangh* log, was more realistic in assessing its agricultural potential, stating that 'the ground is covered with little or no soil, mostly white sand and rocky, and in my opinion not suitable for cultivation'. Torst, throughout his log, expressed similar adverse opinions about other aspects of Terra Australis.

In 1696, much of the island would have been covered by forests of cypress (*Callitris preisii*, or Rottnest Island pine), interspersed with patches of tea-tree (*Melaleuca lanceolata*), cheesewood (*Pittosporum phylliraeoides*), and wattle (*Acacia rostellifera*). Open, low-lying plains would have surrounded parts of the salt lakes. The first botanist to visit the island, Allan Cunningham, landed at Thomson Bay in 1822, where he was impressed by the abundant cypress trees, which had trunks about 3 feet (1 metre) across and 25 feet (8 metres) tall. He said that they gave the vegetation of the island a 'monotonous appearance'.[40] These large cypress trees must have been hundreds of years old, and they no doubt included some that had been seen by Vlamingh 126 years before. Sadly, the sweet-smelling cypress forests have now disappeared from the island because of extensive timber cutting during the early years of colonial occupation, combined with periodic bushfires and the subsequent destruction of new seedlings by quokkas. All that remains of the former cypress forest are a few scattered trees and some recent plantations. However, a number of attractive groves of tea-trees have survived, and an extensive regeneration program, planting cypress and other indigenous trees, is now in progress.[41]

Vlamingh's men went ashore several times from 1 to 4 January in order to collect firewood and fresh herbs, said to have included 'myrrh', 'aloes', and 'incense' (presumably for medicinal purposes and to suppress unpleasant odours on board the ships). Vlamingh himself landed for a second time on the morning of 2 January, and was again enthusiastic about the trees, describing them as 'the finest wood in the world, from which the whole land was filled with a fine pleasant smell' (*Geelvinck* log).

Vlamingh's journal and the *Geelvinck* log show how much the expedition members enjoyed their visit to this delightful island. They must have departed reluctantly, feeling rather apprehensive about going ashore on the mainland coast. Numerous fires could be seen burning there, pointing to the presence of many native people — people whom they had been led to believe were dangerous savages.[42]

A place called Belvedere

SWAN RIVER

On 4 January, the uppersteersman and several other crew from the *Geelvinck* sailed from Rottnest Island to the mainland in the longboat, returning on the same day. Their landing point on the mainland coast is unclear. Vlamingh's journal recorded that:

> In a cove they had seen several small islands [with water] 4 to 5 fathoms [7 to 8.5 metres] deep all around them.[43] They had found sometimes a foul and bad bottom, and sometimes a fine and good one, and after much trouble they had finally gone ashore, where they had found many trees dripping gum, of which they brought some aboard. And since they deemed themselves to be too weak in manpower, they had not dared to risk going any further ashore.

The 'small islands' were probably the chain of rocks and tiny islands that extend between Rottnest and Carnac Islands, and which can be seen readily from a boat when sailing from Rottnest to the mainland. In the one day taken for their trip to the mainland and back, there would not have been enough time to sail or row around the larger Carnac and Garden Islands.

The 'many trees dripping gum' seen close to the coast were probably wattles, most likely belonging to the species *Acacia saligna*,[44] which exude gum when affected by boring insects. Another explanation is that the gum came from marri trees (*Eucalyptus calophylla*), which exude copious quantities of reddish-coloured gum from injuries to their trunks. However, it seems unlikely that the small group would have walked far enough inland to reach a belt of marri trees, because of the perceived danger of attack by native people.

On 5 January, a party of eighty-six officers and men under the command of the skipper of the *Nyptangh*, Gerrit Colaert, and including the bookkeeper (Joannes Bremer) and the uppersteersman of the *Geelvinck* (Michiel Bloem), sailed to the mainland in the galliot *Weseltje*, arriving there at noon. Vlamingh stated in his journal that he had planned to participate in this landing, but had decided 'to postpone going there until I could be informed of the success of this first landing, before making another with more ease'. Of course, Vlamingh was then fifty-six years old, which in those days meant that he was an old man; indeed, only one year later he would be described as 'aged and feeble'.[45] Walking for long distances through the bush was a task for active young men, not for persons of Vlamingh's age, and that was probably an unstated reason why he did not join the exploration party.

Previous authors have suggested that the landing point on 5 January was probably about 2 kilometres south of Cottesloe Beach, where the Vlamingh Memorial was erected in 1973.[46] However, it seems more likely that the party landed at Swanbourne Beach, 4 kilometres to the north (see Figure 11), as its situation is more consistent with information recorded in the *Geelvinck* and *Nyptangh* logs.

In the *Geelvinck* log, it was reported that the men, on walking inland, skirted around an 'inland lake' (probably Lake Claremont, 2 kilometres east of Swanbourne Beach).[47] The party then continued east, observing many signs of the native inhabitants, including several 'huts', said to have resembled those of the Hottentots at the Cape of Good Hope. After continuing on for more

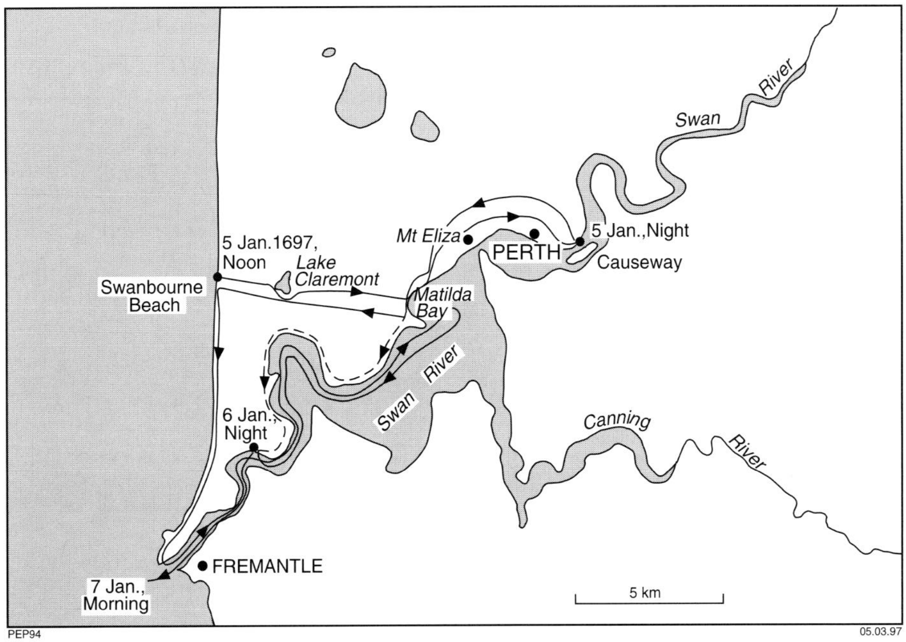

Figure 11. *Map of the Swan River area, showing the approximate routes followed by the party led by Gerrit Colaert on 5–7 January 1697.*

than one and a half hours (perhaps following Aboriginal paths), they came upon a body of water, and realised that it was part of a large river, which was later given the name Swan River (*Swane Rivier*). The party may then have been at Matilda Bay. On the river beach they noticed many human footprints, including those of adults and children.

The *Nyptangh* log was less informative about the course followed, stating only that they marched inland, and after about an hour found 'a small hut quite as bad as those of the Hottentots', then 'somewhat further on a large basin with brackish water [probably Melville Water], which later was found to be a river'. Vlamingh's own account sheds little light on the course followed, stating that 'after marching [to the east] for an hour 'they came to a hill, where having seen a river in a glen they went down to it'. However, as Vlamingh was not a member of the exploration party, his description was necessarily hearsay, and more reliance can be placed on the *Geelvinck* and *Nyptangh* accounts, as they were written by members of the party. Putting those two accounts together, it would seem clear that the walk from the coast to the Swan River probably occupied more than two hours. They could easily have covered the distance from Swanbourne to Matilda Bay (6.5 kilometres) in that time if they walked at a about 3 kilometres an hour, which seems quite feasible. A walking time of more than two hours is supported by Valentyn. He reported that the party arrived on shore at 11 o'clock in the morning and reached the river three hours later.[48]

Following the river inland, the party arrived at the foot of 'a steep area', probably the face of Mount Eliza, where it was no longer possible to walk along the beach. Accordingly, they climbed to the 'high ground' (presumably the present Kings Park), and found a very large tree, described in the *Geelvinck* log as being 3 Dutch fathoms (5 metres) in diameter. This striking tree, which must have been a giant jarrah, marri, or tuart, was described in Vlamingh's journal as 'a big, cool, green tree … which was full of knots from top to bottom and very easy to climb to the top'.

The *Geelvinck* log recorded that it had notches cut into the trunk by which 'people could go up to the top of the tree'. Presumably the cuts were made by an Aborigine seeking to catch possums.[49]

After resting under this great tree and taking refreshments, the party left the 'high ground' and rejoined the river further upstream, perhaps near the present Perth Esplanade, noticing several holes dug there by the native inhabitants to obtain fresh water. The river water at this locality was said to have been 'a little brackish' (Vlamingh journal), 'brackish' (*Geelvinck* log), or 'somewhat brackish' (*Nyptangh* log), whereas nowadays the river near Perth during summer is normally as salty as the sea; it could never be described as 'a little brackish'. Part of the reason for this is that the Swan River's main source, the Avon River, became saline after its catchment was cleared for agriculture, while the Helena River, a major freshwater tributary, has been dammed. Moreover, the bar across the mouth of the river, described in the ships' logs and Vlamingh's journal, was removed in 1895–97 to allow for the development of Fremantle Harbour, thereby resulting in stronger tidal exchange of water between the sea and the estuary.

It also seems likely that at the time of Vlamingh's visit the climate was wetter than it is now, so that fresh water continued flowing into the upper reaches of the river well into summer. This view may be supported by the fact that the 'inland lake' reported by the landing party, thought to have been Lake Claremont, must have contained water on 5 January 1697, whereas in recent times, before the area was cleared for urban development, the lake would dry out completely in summer.

A wetter climate is also suggested by the fact that Hutt Lagoon (450 kilometres north of the Swan River, see Figure 4) still contained water when Vlamingh's men visited it on 23 January 1697 (see page 46), whereas in recent times it would normally have been dry by early summer.

It is also noticeable that Vlamingh often made comments in his journal regarding the favourable climate of Terra Australis. He commonly described the weather as 'very pleasant', 'very clear and pleasant', and 'very fine', never complaining about high temperatures. The only adverse weather was experienced at Shark Bay during the period from 4 to 11 February, when rain and strong winds, associated with summer thunderstorm activity, made conditions unpleasant.

It must be remembered that at the time of Vlamingh's visit, Europe was experiencing the Little Ice Age — a frigid period that lasted for about 500 years up to the mid-nineteenth century. The coldest average temperatures during that period were experienced in the seventeenth century, probably a little earlier than Vlamingh's visit to Terra Australis. The effects of the Little Ice Age on the climate of Australia are not adequately understood, but the reports of Vlamingh's expedition and other data suggest that the south-west area was probably significantly cooler and wetter then than it is now.[50]

As the Dutch party walked eastward beside the Swan River in the late afternoon of 5 January, they could see several plumes of smoke rising in the distance from fires lit by the natives. They must also have observed the Darling Range, although that fact is not recorded in the ships' logs or Vlamingh's journal. However, Witsen referred to the range in his account of the expedition, based on the verbal reports he had received from expedition members.[51]

Five men, including two 'Indian' interpreters, were quietly sent ahead of the rest of the party in an attempt to contact the natives without frightening them away. This tactic proved unsuccessful — no Aborigines were found, even though there were many signs of their recent presence in the area. It was concluded that 'those people, on noticing ours, had withdrawn and hidden' (Vlamingh journal).

The party camped for the night beside the river, carefully posting sentries to guard against attack by the natives. They were then said to have been 2–3 Dutch miles (15–23 kilometres) inland, but if they were near the site of the present Causeway, which seems likely, they were really only about 1.5 Dutch miles (12 kilometres) east of their landing point.

Next morning, the group split into three parties, each of which followed a different route, hoping to 'catch' and 'lay aboard' one of the native inhabitants (*Geelvinck* log and Vlamingh journal). One of the parties was said to have walked 1 Dutch mile (7.4 kilometres) to the east, while the others went north and south.[52] However, although more 'huts' were found, no Aborigines were seen. The men observed various white, green, and grey birds, some of which they shot, including 'cockatoos and parrots with great curved beaks' (Vlamingh journal). They noted that the water level in the river had fallen by a foot since the day before, and realised that this must have been the result of tidal action.

The three parties came together again at a pre-arranged rendezvous, perhaps near Matilda Bay, where the river was said to have been up to 2 Dutch miles (14.8 kilometres) across (*Geelvinck* log). Again, this is an overestimate, as the maximum width of the river, at Melville Water, is about 7 kilometres north–south and 4 kilometres east–west.

At their rendezvous, several men roasted and ate fruit of the Zamia palm (*Macrozamia riedlei)*, which the uppersurgeon of the *Nyptangh* (Mandrop Torst) described as tasting like Dutch broad beans, or, when less ripe, like hazelnuts. However, he recorded that within three hours after eating the nuts, he and the others 'began to vomit so violently that there was hardly any distinction between death and us' (*Nyptangh* log). Of course, the men had been unaware of the extremely poisonous nature of these nuts when untreated.[53]

Some of the party continued walking along the river bank, while others, including the sick people, returned to the boats on the coast. The sick men went aboard the galliot, taking with them a plank from a wrecked ship (possibly the *Vergulde Draeck*) that had been found on the ocean beach. The rest walked south along the coast to find the mouth of the river, reaching it after a march of two and a half hours, according to the *Geelvinck* log, although Vlamingh's journal states that it took only half an hour. This part of the *Geelvinck* log is more likely to be reliable than the equivalent part of Vlamingh's journal, as the log is acknowledged to have been written by a member of the shore party (probably the bookkeeper, Joannes Bremer), whereas Vlamingh had remained on board the ship. A walking time of two and a half hours from the landing point to the mouth of the Swan River would confirm the evidence discussed earlier that the party came ashore at or near Swanbourne Beach.

The longboat kept abreast of the party that was walking along the coast towards the river mouth. On reaching the mouth, the men managed to drag the boat over the rocky bar to enter the river. They then sailed and rowed upstream looking for the other men, and observing many fish and birds, especially black swans. After locating the other party, they spent the night on the river bank, returning to the galliot early the next day, 7 January.

The three ships anchored together off the mouth of the river on 8 and 9 January (see Figures 12 and 13).[54] On the 10th, forty of the men, led by Vlamingh himself, set out to explore the river in two longboats and a pinnace, crossing the rocky bar with some difficulty. Two swivel cannon were mounted on the longboats, no doubt for defence. After sunset, they sailed and rowed up the river in bright moonlight for an estimated 5–6 Dutch miles (37–45 kilometres), until about two o'clock in the morning, when 'as our men felt tired we deemed it necessary to drop our grapnels so as to be more able to go and explore on land the next day' (Vlamingh journal).

The men rowed on from daybreak on 11 January, capturing several black swans and shooting nine or ten others, and seeing many other birds including divers and 'brent geese' as well as schools of fish 'floundering in the water' (*Nyptangh* log). They also thought they 'heard a nightingale sing'.[55] The captive swans were subsequently taken back to the *Geelvinck*, and three survived the

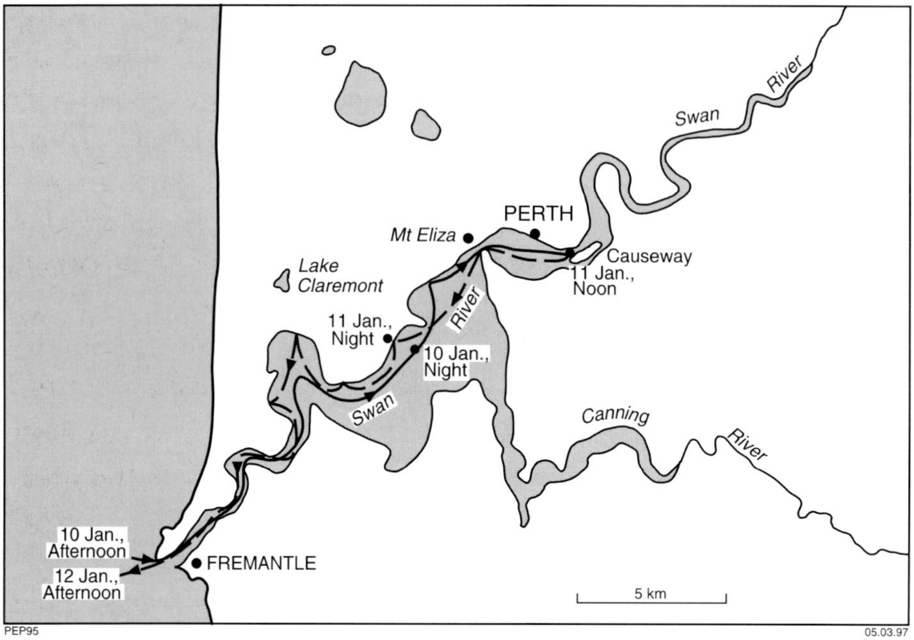

Figure 12. *Map of the Swan River area showing the approximate route followed by the party led by Willem de Vlamingh in three ships' boats on 10–12 January 1697.*

voyage to Batavia, only to die there, one by one, before they could be taken to the Netherlands.

Several of the men went ashore at a place where they might have briefly glimpsed two black people. However, nothing was found other than a few 'huts', fires, and human footprints — no human beings were to be seen. The boats continued up the river to a place where it divided into a number of branches and became too shallow for the boats to proceed further. The river water was said to have been 'still a bit brackish', and several snakes were seen in the water (probably tiger snakes). The men estimated that they were then 9–10 Dutch miles (67–74 kilometres) up the river, according to the *Geelvinck* log and the Vlamingh journal, or 6–7 Dutch miles (44–52 kilometres), according to the *Nyptangh* log. If the first of these estimates is correct, it would seem that the party must have rowed and sailed as far as the Upper Swan area, while the second estimate would point to the Middle Swan area. However, an inspection of Victorszoon's map of the area suggests that they went no further than the site of the present Causeway, where the Aborigines used to ford the river and subsequent explorers often found the water too shallow for boats. If so, the party had travelled less than 3 Dutch miles (22 kilometres) from the river mouth, even allowing for the twisting course required to follow the deeper part of the river channel. This distance is less than a third of the lower estimate given in the *Geelvinck* log and half that of the *Nyptangh* log.

The boats turned back at noon, and the men went ashore at a scenic place, said to have been about 3–4 Dutch miles down the river from their turning point, or using their previous distance estimates, between a third and half of the way back to the river mouth. In his journal, Vlamingh stated that on arriving at this place:

> We climbed first up a little hill, which could properly be given the name of 'Belvedere' since from there a very fine countryside is displayed on all sides, open land mixed with forest. We went in all directions in the hope of meeting with some of the local inhabitants, but our exertions and moves were fruitless, and we were forced, through fatigue and nightfall, to go away from there in our boats, in order to find, close to the river bank, a suitable place to lie down and rest. Having found same, I also wanted to turn this place into a general rendezvous. For this reason I had sentries posted all around to guard us from surprises and dangers which could occur, and since some of the inhabitants, at the sight of our fires, might come close, I ordered the sentries to be very watchful all around.
>
> At 2 o'clock at night, upon observing the glow of several fires, we boarded very softly and rowed noiselessly towards that place, and when we were close enough we despatched eight well-armed men with two Indians to try to catch one of the inhabitants. They went straight to these fires, but when they arrived there they neither found nor saw anybody.
>
> On the 12th, from daybreak, we rowed to another place where we set foot ashore and found several more huts, one foot tall on one side, and two on the other. Inside them were pieces of bark from trees [presumably

Figure 13. *Sketch of Vlamingh's three vessels, the* Weseltje, *the* Geelvinck, *and the* Nyptangh *(left to right), lying at anchor off the mouth of the Swan River. Two ships' boats are seen about to enter the river, on which a number of black swans are swimming.*
From Valentyn (1726), courtesy of the Rare Books Section, Fisher Library, University of Sydney

paperbark for use as bedding], and on the outside they were covered with fine white reeds.

After progressing some distance we saw gum on nearly all of the trees, of which we took enough to serve as samples and evidence, and since I saw that our exertions would not yield further finds, I resolved to go back on board, which I did without any further consultation, so that we arrived on board all together in the afternoon.

A sample of gum (probably from either an *Acacia* or a marri tree) was taken back to the Netherlands and delivered to Nicolaes Witsen. He reported that the trees 'produce a great deal of reddish-brown-coloured gum which drips from the heart of the tree and falls down in drops like little balls, a sample which has been brought here, being the bark of a tree to which this resin or gum is still attached'.[56]

It is not possible to be sure of the location of 'Belvedere' (meaning 'beautiful view'), as there are several places beside the Swan River that might qualify. Perhaps 'Belvedere' is situated somewhere along the present Crawley–Dalkeith foreshore, overlooking the nearby river flats. Alternatively, the 'little hill' might be Butler's Hump in the Keane Point Reserve, which provides a view of the attractive environs of Peppermint Grove and Mosman Bay.[57] The 'open land' may have consisted of areas that were burnt regularly by the Aborigines.

Vlamingh must have decided on 12 January that his quest to 'catch a South Lander' was hopeless, and that little would be achieved by further searching in the area. The Aborigines must have fled, terrified by the strange white beings who crept up on their camps during the day and late at night. They no doubt believed that the whites were *jangga*, spirits of dead Aborigines who had returned to their old camping sites from *kurranup*, the home of the dead over the great sea.[58]

In deciding to go back to the ships 'without further consultation', Vlamingh must have been aware that several members of his party were not happy with that decision, as they wanted to explore further inland. After returning to the Netherlands, some of them spoke to Nicolaes Witsen, who recorded that:

> He [Vlamingh] discovered a high mountain range [the Darling Range] and saw a great many footprints, both of grown-up people and children, pointed in that direction, and he also thought to see much smoke below that mountain range, so that there will probably be a great number of people there at the foot. He considered walking in that direction, but bearing in mind that they were already eight to ten miles inland and that the ships were lying at anchor far out to sea, and being worried that he might be surrounded by a great crowd of people, he retreated, although

> many in his company would have ventured the approach, as they reported to me orally, confident in their firearms and the timid nature of the inhabitants.[59]

Because of the abundance of swans, the river explored on the expedition was named *Swane Rivier* (Swan River) in the descriptive text accompanying the map of the *Zuyd Landt* (South Land) drawn by Victor Victorszoon. However, it seems likely that Vlamingh himself had originally intended naming it the 'Witsen River', after Nicolaes Witsen. Witsen was clearly led to believe that this had been Vlamingh's intention, stating in his 1705 publication that 'At the time of its discovery, by the commodore Willem de Vlaming, this river was named the Witsen River after me, as may be seen in the charts drawn by him on the spot and now in my keeping'.[60] However, those charts have since been lost, and it was Victorszoon's name, 'Swan River' that would prevail.

Most expedition members probably had few regrets about leaving the Swan River area, complaining that the bush there was 'full of prickles and thorns' and 'millions of flies'. Mandrop Torst, in the *Nyptangh* log, reported that 'the best part of it is that no vermin are found there, but by day the flies are a terrible torment'. Moreover, Torst and the other men who had eaten fruit of the Zamia palm had survived an ordeal they would never forget.

Vlamingh himself, however, made few negative comments about the Swan River and Rottnest areas, other than expressing disappointment at his lack of success in catching an Aborigine. Indeed, he made favourable remarks in his journal regarding the beautiful view from 'Belvedere', the many attractions of that 'paradise on earth', Rottnest Island, and the 'very pleasant weather' experienced during their visit.

A barren land

Swan River to Shark Bay

On 13 January, the ships left their anchorage near the river mouth and sailed north, close to the coast. On the 15th, while passing near the site of the present town of Jurien, expedition members saw a great deal of smoke and vapour rising from the land. Thinking that this indicated the presence of native people, Vlamingh decided to send a party ashore in an endeavour to contact them. The party returned in the evening to report that the land consisted of barren sand dunes, which were 'not fit for animals, let alone humans, to inhabit' (*Nyptangh* journal). However, they did see footprints of people and very large birds that they thought might have been cassowaries (no doubt they were emus).

In the evening, fires could be seen all along the coast, so two boats were sent to land again next day. The men walked inland, again hoping to contact the native people. They found a freshwater marsh with a 'hut' nearby, and saw human footprints and tracks of 'tigers' (presumably dingoes) and other wild animals. Some very large fires were burning in the area, but, as usual, the people had disappeared.

Throughout the ships' logs and Vlamingh's journal, there are similar references to bushfires extending along much of the coast. These observations point to the Aboriginal practice of burning the bush extensively during summer, when the vegetation was driest and fires spread most widely. Although the effects of such fires would be very destructive from today's perspective, they were beneficial to the Aborigines. Widespread burning 'opened up' the country, allowing better access for hunting, and the rains that followed in autumn and winter stimulated growth of new vegetation, thereby attracting kangaroos, other marsupials, and birds to the burnt areas. These fire practices had been in place for tens of thousands of years, and the biological equilibrium that existed prior to European colonisation depended on regular burning of the bush by the Aborigines.[61]

On 16 January, the wind turned around to the north, and over the next two days the ships drifted south for about 60 kilometres (see Figure 4). On the 17th, they noted that the mainland had high sand dunes, 'like those of the Vlie' (Vlamingh's birthplace), with two distinct hills resembling small forts at the crests of the high land. These were sketched by Victor Victorszoon (see Figure 7), and can be identified today as North Hummock and South Hummock — two prominent mounds of sand overlying rough limestone near The Pinnacles, south-east of Cervantes.[62] A party was sent inland to explore these features, and returned to report (*Geelvinck* log) that they had walked through 'barren land, without trees' and climbed 'a high stony mountain, which we mistook for a fortress and which resembles Dover Castle' (presumably North Hummock). From there they had observed 'an infertile desert' extending 'as far as the eye could see' (Vlamingh journal). On the way back to the ship they visited a small island which was 'full of young seagulls', and took away about 100 of them, no doubt to eat.

Today, much of the country around North and South Hummocks, and parts of those two hills, are covered with thick coastal vegetation, consisting of a variety of shrubs and low trees. From Victorszoon's sketch of the locality (see Figure 7), and descriptions given in the *Geelvinck* log and Vlamingh's journal, it seems clear that this area was not nearly as thickly vegetated at the time of the Dutchmen's visit. The explanation is probably that the Aborigines living in the area ensured that the coastal scrub was burnt regularly, keeping the vegetation lower and less dense than it is today. This regular burning by the Aborigines probably gave a desert-like appearance to the area.

On 18 January, a fresh wind was blowing from the west, and they resumed their northerly voyage. Another party was sent ashore on 20 January, again near the present site of Jurien, returning after observing 'a real desert, treeless and without any other vegetation' (Vlamingh journal).

The ships sailed north, past Beagle Island, and boats were sent ashore from the *Nyptangh* on 21 and 22 January. On returning, the men reported that they had found a pool of salty water about 200 paces from the beach, and had walked inland along it for about a quarter of an hour, finding that it was fairly deep in its centre and 'well stocked with fish'. This body of water may have been the estuary at the mouth of the Greenough River. Again they noticed 'tracks of tigers and other wild beasts' (Vlamingh journal).

On 22 and 23 January, the ships sailed through the wide passage that is known today as Geelvink Channel, which extends between the mainland coast and the islands and reefs of the Houtman Abrolhos. While sailing near the present site of Geraldton, they noticed a high mountain about four Dutch miles (30 kilometres) inland, 'having the same appearance as the Table Mountain at the Cape of Good Hope' (Vlamingh journal). This was probably Mount Fairfax, at the southern end of the Moresby Range.

The first definite sighting of Aborigines during the expedition was made on 23 January, just north of the present township of Gregory. The crews of two ships' boats saw ten naked black people, three or four on the beach and the others on the dunes. According to Vlamingh's journal:

> While these blacks remained long on the shore they made signals to our people with their feet and hands. But since there was a great surf which drew the boat along towards the shore, this stopped them from landing at once. And when they got there at last, before they had put a foot

> ashore, these black coastal dwellers fled at such an extraordinary speed that it was impossible for our people to follow them.

The men returned to their boat, and one Dutch mile (7.4 kilometres) to the south they were again able to land safely. On walking inland, they saw what is now known as Hutt Lagoon, describing it as a 'blood red' body of water, and commenting that it was possible to collect 'as much salt as one would like' (*Geelvinck* log). In his journal, Vlamingh deduced that sea water must sweep over the narrow strip of coast into the lagoon during very high tides, and that 'through the power of the sun' the water in the lagoon would evaporate, leaving salt. Indeed, this is a valid explanation of the way in which salt has been precipitated in the lagoon.[63]

After the visit to Hutt Lagoon, the three ships sailed on in the afternoon of 23 January, and towards the evening 'a very steep coast was seen, about two [Dutch] miles long, which began to rise in the north-north-east and further east-north-east' (*Geelvinck* log). At the end of these cliffs they reached 'a brave inlet', defined on its south side by *Roode Houk* (Red Bluff), and the ships came to anchor in what is now known as Gantheaume Bay, a short distance south-west of the present township of Kalbarri. Next day, 24 January, the ships' council resolved that 'the boats should sound the inlet today' and that parties of men should be sent ashore the following morning to explore the countryside, 'which appears so pleasing in this place, through its hills, valleys, and plains' (Vlamingh journal).

On 25 January, a party of forty-six well-armed men was landed to explore the hinterland, with enough food for eight days, while a separate party of fourteen men, including the uppersteersman and bookkeeper of the *Geelvinck*, proceeded elsewhere. The smaller party returned later that day to report having found a hole containing fresh water, with ten to twelve footpaths leading to it, and five 'huts', but once again no human beings had been seen. A later report indicated that this waterhole was about an hour's walk from the beach, and that one of the 'huts' was made of clay, with a roof 'sloping down on two sides'.[64] The party also advised that they had found a place beside a marsh (no doubt the lower part of Wittecarra Gully) that was 'full of frogs' (Vlamingh journal).

The larger party marched towards a 'mountain range' (the elevated land east of Wittecarra Gully). As a member of this party, Mandrop Torst reported in the *Nyptangh* log that:

> [We] sometimes had to rest through fatigue caused by the heat of the sun and the heavy going through thick scrub, until we came to the mountain range, where we camped. But if the march had been hard, the greatest grief struck us now, for finding no water, we thought we would perish from thirst. We could see our ships clearly from here and wished a thousand times to be back on board.

However, three men sent to look for water eventually located a slightly brackish supply in a 'great pit', one hour's march from the 'mountain range'. Further exploration in the vicinity disclosed only sand, 'thick and thorny' scrub, and a 'hut'. They were very glad to return to the ships, with 'the men complaining bitterly of sore eyes' (*Nyptangh* log).

The landing parties must have gone ashore either at the little beach immediately north of Red Bluff or a second beach nearby at the mouth of Wittecarra Gully. The waterhole found by the smaller party, about an hour's walk from the beach, must be Wittecarra Spring (see Figure 14). This was probably also the source of fresh water discovered by Francisco Pelsaert's party in 1629, when he marooned two convicted criminals from the *Batavia*, Wouter Loos and Jan Pelgrom.[65] Until very recently, this beautiful spring could be seen much as it was in 1697, but in 1996 it became dry due to high rates of extraction of groundwater from a nearby bore.[66]

The second water supply found by the larger party (the 'great pit') might have been Gabba Gabba Spring, or alternatively some other waterhole excavated by Aborigines that is now filled in and forgotten.

The *Geelvinck* and *Nyptangh* logs and Vlamingh's journal indicate that the landing parties never saw the Murchison River, which flows into the northern part of Gantheaume Bay. The river must have been obscured from

Figure 14. *Wittecarra Spring, 3.5 kilometres from the mouth of Wittecarra Gully, in October 1994. This spring was found by one of Vlamingh's parties in 1697, and probably also by Pelsaert's party in 1629.*
Photograph by the author

their view by dense scrub and the nature of the topography. The first recorded discovery of this river was by George Grey, 142 years later, in April 1839. He came ashore about two kilometres north of the Dutchmen's landing point (see Figure 15), at a place where a continuous rock platform in front of the beach is difficult to see from the sea when the tide and waves are high. Grey failed to discern this rocky platform until it was too late, and his boats were irreparably damaged while coming ashore. He and his party were left with no option but to walk 500 kilometres through unknown country to Perth, a dreadful ordeal.[67]

On 28 January 1697, Vlamingh's ships weighed anchor and sailed north-west from Gantheaume Bay, below the towering Zuytdorp Cliffs. The

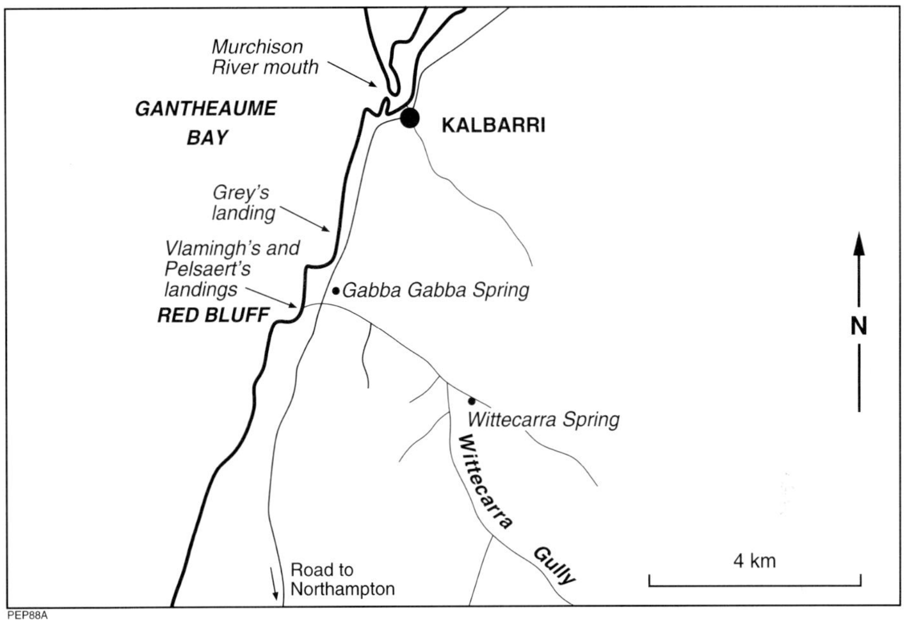

Figure 15. *Map of the area around Kalbarri and Wittecarra Gully, showing the locations of Wittecarra Spring and Gabba Gabba Spring, and the probable landing points of Pelsaert (1629), Vlamingh (1697), and Grey (1839).*

following morning, while sailing about half a Dutch mile (3.7 kilometres) offshore, these cliffs were described (in the *Geelvinck* log) as:

> High steep land and very precipitous … altogether high and bare, without any green thing on it and very steep up to the top … without any beach at its foot, [and] with heavy surf that is a wonder to see.

In his journal, Vlamingh described the cliffs in even more graphic terms:

> The land here appears very bleak, and so abrupt as if the coast had been chopped off with an axe, which makes it almost impossible to land. The waves break here with so great a fury that one should say that everything around must shake and become dismembered, which appears to us a truly

> terrible sight. Nevertheless, we found it rather difficult to drag ourselves away from this evil place.

Vlamingh and his men must have felt mesmerised by the frightening scene. Little did they realise that only fifteen years later this place would become the graveyard of a great ship of the Dutch East India Company, the *Zuytdorp*, which crashed into the base of the cliffs during the winter of 1712. I am sure that Vlamingh and his men could never have envisaged that anyone might live through a shipwreck at this fearsome locality, but some of the *Zuytdorp* crew *did* manage to survive and scramble ashore, although none ever returned to civilisation to tell their story.[68]

A story of two plates

Dirk Hartog Island

On 30 January, Vlamingh's ships came to anchor in what is now known as South Passage, between the southern end of Dirk Hartog Island and Steep Point (*Steyle Houck* on Victorszoon's map). Boats were despatched over the next four days to sail around the island. On the morning of 2 February, men in the *Geelvinck*'s pinnace went ashore at the north end of the island, at what later became known as Cape Inscription (see Figures 5 and 7). They climbed the cliff there, finding an oak post fixed in a cleft in the limestone. Beside it lay a pewter plate, evidently fallen from the post, with an inscription recording the arrival of Dirk Hartog at this place more than eighty years before. This inscription may be translated as follows:

> 1616, the 25th October, is here arrived the ship the *Eendracht* of Amsterdam, the upper-merchant Gillis Miebais of Liege, skipper Dirck Hatichs of Amsterdam; the 27th ditto set sail again for Bantam, the under-merchant Jan Stins, the uppersteersman Pieter Dookes van Bill, Anno 1616.

Figure 16. *Aerial view of Cape Inscription, Dirk Hartog Island, showing the lighthouse, ruins of the lighthouse-keepers' quarters, and the place where Dirk Hartog and Vlamingh left their posts and plates (indicated with an arrow).*
Photograph by the author

The text was deeply engraved from the beginning to the word 'Bantam', but the rest was only lightly scratched in double lines on the surface of the plate, probably using the point of a knife. Perhaps the ship had to leave before the engraving could be finished properly, or Jan Stins and Pieter Dookes van Bill managed to have their names added shortly before the plate was put in place — we will never be sure of what happened. No facsimile copy of the inscription has previously been published, and as the plate has deteriorated badly over the past 380 years, most of the lightly scratched part of the inscription has now disappeared (see Figure 18).[69]

Figure 17. *The cleft in the rock on the cliff top at Cape Inscription, where Dirk Hartog and Willem de Vlamingh left their posts and plates, showing two jarrah posts placed there by the Western Australian Government in 1908.*

Photograph by the author

The Hartog plate was taken on board the *Geelvinck*, and Vlamingh recorded in his journal that from it 'I learnt that I was truly in the Bay of Dirk Hartog, of which I have often heard tell'. Vlamingh took the plate with him to Batavia, realising its great historic importance, and from there it was transferred first to the VOC's archives in Amsterdam, and later to the Rijksmuseum in Amsterdam. It apparently disappeared there for nearly 200 years, but was eventually relocated in 1899, when a Dutch scholar, J.F.L. de Balbian Verster, found it on display in the museum, incorrectly labelled as having come from the shores of the Straits of Magellan in South America.[70]

Unfortunately, the plate had already deteriorated badly by then, as a result of advanced 'pewter disease' (oxidation), and it disintegrated further over the next sixty years, before appropriate conservation measures were taken.

On 4 February, Vlamingh's ships were moved to anchor at Dirk Hartog's roadstead, beside the north end of the island. Expedition members explored parts of Shark Bay for several days without finding anything of particular note, other than many turtles, some of which they slaughtered or captured for meat. At the place near Cape Inscription now known as Turtle Bay, it was recorded in the *Geelvinck* log that 'at night one can catch and turn over as many turtles as one likes, and get as many turtle eggs as one desires'. Victor Victorszoon depicted the east side of the island in one of his watercolour sketches, showing Turtle Bay and the site at Cape Inscription where the plate was found (see Figure 7). Beside the sketch of the post and plate is written (in Dutch) 'here the pewter plate was found'.

Hundreds of turtles still come ashore at Turtle Bay during the egg-laying season (November–February), each turtle laying very large numbers of eggs. I spent a night and morning at Turtle Bay in early February 1955, witnessing many turtles dragging themselves ashore, excavating their nests, and laying eggs, and seeing hundreds of newly hatched turtles scurrying down the beach to the water. This impressive scene was no doubt almost the same as that witnessed by Vlamingh's party in the same month of 1697.

The Dutch seamen also enjoyed other types of fishing in Shark Bay. One notable entry, in the log of the *Nyptangh* for 7 February, recorded that they 'caught a fish of immense size, which twenty-four of us ate … and there remained enough for thirty more persons'. On another day, they 'caught three great sharks, one of which had thirteen young with it [each] in size like a large pike'.

It is difficult to accurately deduce the courses followed by Vlamingh and his men when they explored Shark Bay in the ships' boats from 4 to 11 February. The weather was severe during most of that time, with strong winds, thunder and lightning, and heavy rain — the first rain and adverse weather they had experienced on the Australian coast. On the 7th, when the wind blew very strongly, Vlamingh observed that they had been subjected to 'as much of it as we could bear'. Under such trying conditions they must have had little idea of the courses they had followed, and this probably explains why it is hard to interpret their logs, and why the Shark Bay area is one of the least reliable parts of Victorszoon's map. The map does not show the presence of Peron Peninsula, which divides Shark Bay into two major elements, although Cape Peron itself appears to be represented as a narrow island.

Vlamingh returned to the *Geelvinck* on 11 February. He learned that during his absence, on 7 February (the day when he had complained about the wind), boats had gone to the island to explore and take measurements. The last boat, with the skipper of the galliot (Vlamingh's son, Cornelis) on board, had been overturned by a sudden gust of wind. This misfortune was noticed by people on the ships lying at anchor, and they immediately fired three cannon and raised a flag at half-mast as a signal to the men on the island that a calamity had occurred. Those men immediately rowed 'with all their strength' towards the overturned boat (Vlamingh journal). However, the galliot weighed anchor and arrived at the spot before the boats, saving all the men except for a carpenter, who drowned after becoming entangled in a sounding line.

Vlamingh decided to replace Dirk Hartog's plate with one of his own. Consequently, on 11 February 1697, he had a pewter plate of the same size hammered flat and inscribed with the text of Dirk Hartog's original inscription, plus a record of his own visit (see Figure 19). On 11 or 12 February, the plate was nailed on a new post, hewn from the trunk of a cypress tree from Rottnest Island, and this was placed in the rock cleft beside what remained of Dirk Hartog's original post.

The inscription recording Vlamingh's visit may be translated as:

> 1697, the 4th of February, is here arrived the ship the *Geelvinck* of Amsterdam, the commodore and skipper Willem de Vlamingh of Vlieland, assistant Joannes Bremer of Copenhagen, uppersteersman Michiel Bloem of the bishopric Bremen, the hooker the *Nyptangh*, skipper Gerrit Colaert of Amsterdam, assistant Theodoris Heirmans of ditto, uppersteersman Gerrit Gerritsen of Bremen, the galliot the *Weseltje*, commander Cornelis de Vlamingh of Vlieland, steersman Coert Gerritsen of Bremen, and sailed from here with our fleet the [12th ditto] also the South Land further to explore, and bound for Batavia.[71]

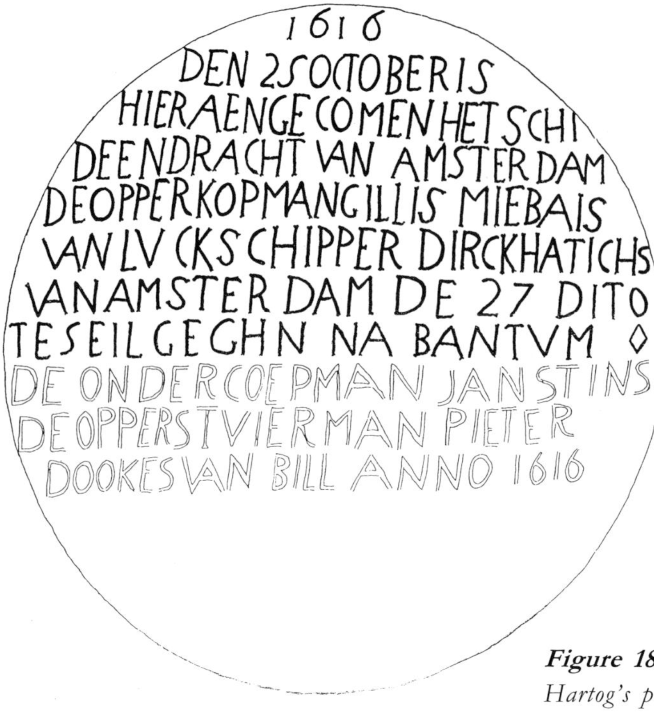

Figure 18. *The inscription on Dirk Hartog's plate: an interpreted facsimile by the author.*

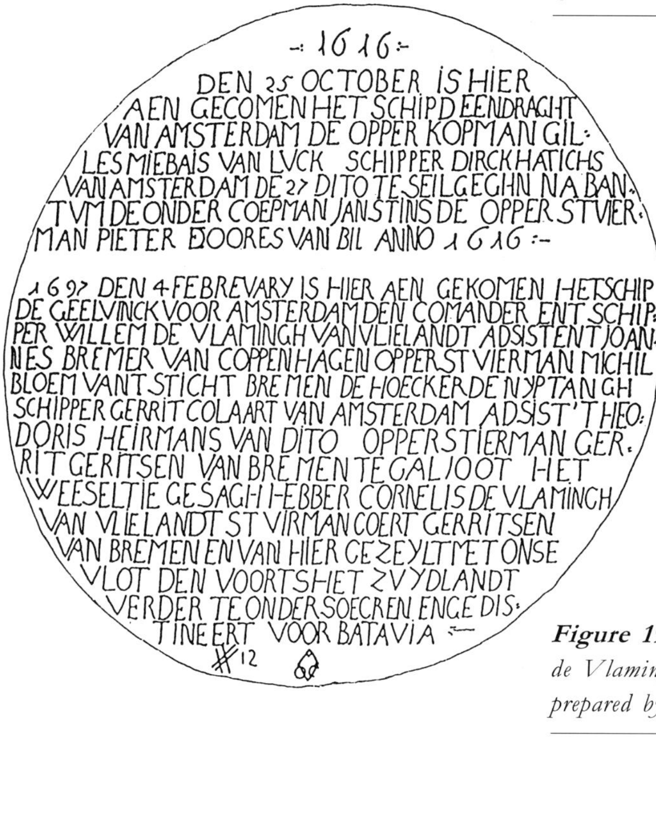

Figure 19. *The inscription on Willem de Vlamingh's plate: from a facsimile prepared by Louis de Freycinet in 1825.*

Vlamingh's plate remained untouched at Cape Inscription for 104 years.[72] It was found in July 1801 by a sailor from the French vessel *Naturaliste* (one of the two ships of the Baudin expedition, the other being the *Géographe*), who had climbed the cliff at this same place.[73] The plate was lying partly covered by sand at the foot of a weathered post (Willem de Vlamingh's), which was standing beside a second more severely decayed post (Dirk Hartog's). The skipper of the *Naturaliste*, Baron Emanuel Hamelin, believed that it would be sacrilege to remove the plate, so he had it nailed to a new post, which was placed in the limestone fissure alongside Vlamingh's original post.

Hamelin also left a lead plaque, inscribed with a record of his own visit, fixed to another post at a prominent headland on the north-east side of the island. However, the precise locality is not identified in Hamelin's journal, and no-one has ever reported finding the French plaque. Perhaps it is still lying buried below sand at the headland (probably Cape Levillian) where it was left in 1801.

One of Hamelin's young officers, Louis de Freycinet, was not happy with the decision to leave the historic Vlamingh plate on the island. Seventeen years later, as commander of the *Uranie*, he sailed across the Indian Ocean to Cape Inscription with the intention of recovering the plate. He found that it had again fallen from its post. In his journal, Freycinet recorded that he 'did not have the same scruples' as Hamelin, and rather than allowing the plate to be 'removed and ruined by some careless seaman', he 'deemed its natural place to be in one of those great scholarly and scientific storehouses which provide historians with such rich and precious documents'. He removed the Vlamingh plate on 13 September 1818 and took it on board the *Uranie*. Unfortunately, the ship was wrecked on the Falkland Islands during the voyage back to France, but Freycinet managed to retrieve the precious plate and take it with him to France in another ship.

Freycinet delivered the Vlamingh plate to the collection of the Académie Royale des Inscriptions et Belles Lettres (Royal Academy of Inscriptions and Elegant Literature) in Paris.[74] The curators there apparently did not treasure the plate in the way that Freycinet had expected, as it disappeared in the

Academy's collection for more than a century. Despite several requests from Australia and consequent searches of the collection, the plate could not be found, and it was deemed to have been lost forever.[75] However, it was eventually rediscovered in 1940, during the darkest days of World War II, by François Renié. He found it lying in a box in the basement of the Academy's library, among miscellaneous material of little value.[76] After the war, the Australian Government renewed representations to France to have the plate returned to Australia, and in 1947, it was graciously presented to the Australian people by the French Government, in recognition of Australia's role as an ally of France during two world wars. The plate now features prominently among the most valued exhibits of the Western Australian Maritime Museum.

The Dutch Government has declined to follow the French example by returning the even more historic Dirk Hartog plate to Australia. However, the Dutch point of view can be readily understood, as the plate is such an important record of their role in the European discovery of the fifth continent, Australia. A replica of the Dirk Hartog plate (in its present deteriorated condition) is on display at the Western Australian Maritime Museum, together with the original Vlamingh plate.

In 1822, Phillip Parker King, the first Australian-born navigator to chart the Western Australian coast, anchored off Cape Inscription.[77] He knew that Hamelin had left the Vlamingh plate behind in 1801, but did not know that it had been taken away by Louis de Freycinet in 1818. King could see two posts standing on the cliff top as he approached the Cape, and climbed the cliff expectantly, only to find, to his disappointment, that the plate had gone. To mark his visit, King left his name on the Hamelin post, together with that of one of his officers, John Septimus Roe, using rows of nails to mark out the letters.

The Vlamingh and Hamelin posts remained standing in the rock crevice at Cape Inscription until 1907, when they were found during survey work for construction of the nearby lighthouse and its associated quarters. The Vlamingh post, of Rottnest Island cypress, had been at the site for 213 years. Only two small splinters of the post, less than 60 cm long, could be seen protruding above the cleft in the rock. The rest of the post, squared and chamfered and about 1.7 m long, had slid down the cleft, where it remained surprisingly well preserved.[78] This post and the Hamelin post can now be seen in the Western Australian Maritime Museum.

In 1908, two jarrah posts were placed in the rock crevice at Cape Inscription by the Western Australian Government (see Figure 17). Small brass plaques were attached to the posts, but were removed in about 1938, because of fears that the cliff was in imminent danger of collapse.[79] In March 1998 the Western Australian Maritime Museum replaced the two jarrah posts with new posts of oak and Rottnest Island cypress, on which it is intended to attach reproductions of the Hartog and Vlamingh plates.

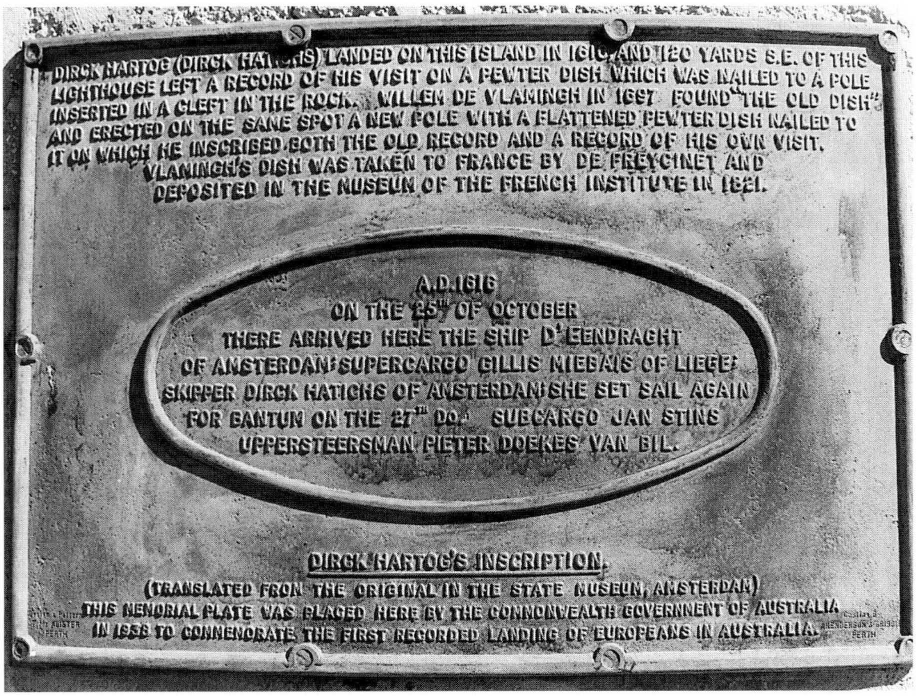

Figure 20. *Bronze plaque placed on the Cape Inscription Lighthouse by the Commonwealth Government in 1938 to commemorate the visit to this locality by Dirk Hartog on 25 October 1616.*

Photograph by the author

Two large brass plaques have been fixed to the wall of the Cape Inscription Lighthouse to commemorate the visits by the two famous Dutch mariners. The first, in honour of Dirk Hartog and commemorating the first proved landing of Europeans in Australia, was placed there by the Commonwealth Government in 1938.[80] The second, marking the tricentennial of Vlamingh's visit, was unveiled by the Premier of Western Australia, Richard Court, on 12 February 1997.

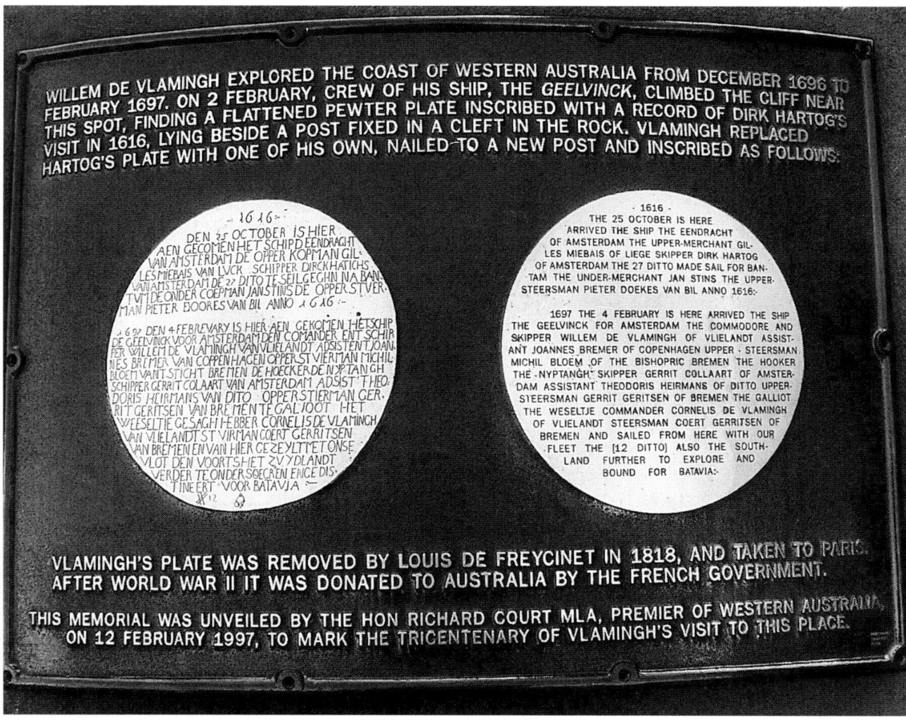

Figure 21. *Bronze plaque mounted on the Cape Inscription Lighthouse, unveiled on 12 February 1997 by the Premier of Western Australia, Richard Court, to commemorate the tricentennial of Willem de Vlamingh's visit.*

Photograph by the author

Farewell to Terra Australis

Shark Bay to North West Cape

Vlamingh's three ships set sail from Dirk Hartog's roadstead shortly before noon on 12 February 1697, heading 'north-north-east and north along the shore' (*Nyptangh* log). Presumably, 'the shore' refers to the coasts of Dorre and Bernier Islands, which are shown on Victorszoon's map but are not mentioned in the ships' logs or Vlamingh's journal. The course followed on 12 February and the following four days in the Shark Bay area is very hard to interpret, especially as there are some significant differences between the accounts given in the two ships' logs and Vlamingh's journal. However, a possible course for this period is shown on the map (see Figure 4).

After leaving Shark Bay, during the night of 17 February, many 'gleams of fire' could be seen on the land (Vlamingh journal) and two days later 'several fires and plumes of smoke' were observed, testifying to the presence of many native people along this stretch of coast. The ships anchored in the bay between what is now known as Vlamingh Head and North West Cape on 19 February, and a party was sent ashore. Next day, a 'great bight' (Exmouth Gulf) came into view, and the ships were brought to anchor beside some small islands (the Muiron Islands), north-east of the Cape.

Vlamingh called meetings of the ships' council on the evening of 20 February and again early in the morning of the following day. A decision was reached to cease exploration of the South Land, as they had already seen enough of this desolate coast. At noon on 21 February, the three ships set sail for Batavia, and Mandrop Torst recorded (in the *Nyptangh* journal) that the *Geelvinck* fired five guns, and the *Nyptangh* three, 'to farewell the miserable South Land'.

Vlamingh must have been anxious about the way in which the directors of the VOC might judge his performance on the expedition. In his journal for the day of departure from Terra Australis, he sought to explain his actions:

> I then gave the signal to weigh anchor and to leave this South Land for good, because the monsoons were over and I concluded that I had done enough in the interests of the company and its directors, my lords and masters, even hoping that they would continue to employ me, and that I could still convince them that I am a very assiduous and devoted servant. I also concluded that it was useless to spend more time exploring and reconnoitring the Terra Australis, believing that I had obtained sufficient information to pass on to them to enable them to send another expedition to this coast, without holding any fear for their vessels, which is a rather important indication of the success of my voyage. Finally, I believed that I had sufficiently fulfilled my duty to my employers, and even to myself, since I had followed and respected the orders and instructions I had been given, such as the making of charts of all these coasts and examining them, as well as my logs and manuscripts, from which it could be seen that, regardless of perils and dangers, I had done everything possible to explore this Terra Australis and its surroundings.

A sad conclusion

Batavia to the Netherlands

After stopping briefly at Christmas Island (then referred to as 'Mony' or 'Moni'), the ships arrived separately at Batavia between 17 and 20 March 1697. Vlamingh recorded in his journal that on arrival at Batavia, he went ashore to make his report to the governor-general and 'ascertain what I should do next according to the orders of my masters'.

Vlamingh remained in the Indies for nearly a year. His health apparently deteriorated during that period, so that he was said to have been 'aged and feeble' by the time he left Batavia for the Netherlands on 3 February 1698. He sailed from Batavia as commodore of the ship *Gent*, and was still on board that ship when it left the Cape of Good Hope on 8 May 1698. His subsequent fate is unknown. No further references to him have been found in company documents or other contemporary records, and it seems possible that he died before arriving home.[81]

During the return voyage, Vlamingh must have felt very apprehensive about the way in which he would be received in the Netherlands, especially if he had received warning of the views of Nicolaes Witsen, one of the most powerful directors of the company. Witsen had already held discussions with expedition members who had returned to the Netherlands before Vlamingh, some of whom must have been critical of their former commodore. From what those men told him, Witsen wrote about the expedition in these terms:

> In short, little has been gained, because the commodore, overmuch inclined to drinking, has not gone much ashore, and has stayed nowhere longer than three days, contrary to his instructions drawn up by myself [which] ordered him to remain long in one place, but he has wasted his time at the Cape with feasting and merrymaking, which grieves me. There are already many people here who have been on the voyage, but the skipper himself is expected in autumn.[82]

Poor Willem de Vlamingh! From these comments, there can be little doubt that he would have received a hostile reception from his 'lords and masters' if he did manage to reach the Netherlands alive. It seems unlikely that they would have accepted his view that he was 'a very assiduous and devoted servant' who should be retained in VOC employment. Whether he died before reaching home, retired because of ill health, or was dismissed from company employment, it must have been a sad ending to his distinguished career.

No profit to the company

RESULTS OF THE VOYAGE

Nicolaes Witsen, who had done so much to promote the expedition, certainly had reason to be disappointed from a purely economic viewpoint about the results of the voyage. In one letter, written in 1698, he said that:

> Nothing has been discovered which can be in any way serviceable to the company. The soil of this country has been found very barren, and as a desert.[83]

Nevertheless, Witsen did eventually find a lot of interest in the journals, maps, and verbal accounts of the voyage, as outlined in various letters he wrote, and it seems that his views on Vlamingh softened with time. Moreover, he was prepared to acknowledge that 'fine forest … and pleasant trees' had been found in the southern part of the area explored, even though the rest was very arid. He stated that:

> From thirty one degrees up to the twentieth degree this South Land or Hollandia Nova, otherwise the Land of the Eendracht, named thus after the Dutch ship which discovered this land in the year 1616, has a desolate and barren sea coast, the further north the more infertile and arid, there being fine forest at thirty-one degrees [the Swan River area], and pleasant trees, but little population close to the coast. The reason is perhaps that there is a lack of drinking water there and the region is not fertile enough to sustain life. Throughout the stretch of coast along which Vlaming sailed he found not a single vessel, so that it seems that the inhabitants there have no knowledge of ship-building.[84]

Most of the places visited by Vlamingh's expedition and earlier Dutch visitors still support little or no agriculture, although the Swan River is now the site of Perth, and the 'fine forest' seen by the Dutch in that area became the basis of a timber industry after British colonisation. The coast from Jurien to south of Cervantes, where Vlamingh sent several parties ashore, is particularly infertile, with little or no potential for agriculture. Consequently, after receiving reports of such inhospitable country, it is not surprising that the Dutch East India Company made no attempt to colonise Western Australia.

Thus, Vlamingh's expedition achieved little of value to the company. Moreover, no evidence was uncovered regarding the fate of the *Ridderschap van Holland*, nor were any castaways found from that vessel or the *Vergulde Draeck*. However, the expedition was certainly very important from the viewpoints of global exploration in the late seventeenth century and later colonisation of this part of the continent by the British. It provided the first accurate map of the west coast of Australia, and resulted in the first recorded discovery and/or naming of important geographic features, especially the Swan River, Rottnest Island, and Dirk Hartog Island.

Vlamingh's discovery of the Swan River was particularly significant.[85] His map of the area became known to other European nations, and over the next 125 years it was explored by several British and French expeditions. The British eventually selected this place as the site for their first settlement on the west coast of Australia — the Swan River Colony, founded at Perth in 1829 —

thereby reinforcing Great Britain's claim to sovereignty over the whole of the Australian continent.

Willem de Vlamingh himself remains a rather shadowy figure, as little is known about his personal life. His journal, however, does shed some light on his character and abilities. He was clearly an excellent seaman and navigator, and at the same time he seems to have been a perceptive, visionary, and very human individual. His favourable comments on Rottnest Island and on aspects of the Swan River area will strike a chord with modern Western Australians, contrasting with the negative opinions expressed by other early Dutch visitors. Indeed, many people today continue to share Vlamingh's romantic opinions of Rottnest Island, the beauty of the Swan River, and the fine climate of south-western Australia.

Epilogue

300 YEARS LATER

At the beginning of this book, I mentioned the celebrations that were held in 1996–97 to commemorate the tricentennial of Willem de Vlamingh's visit to Western Australia. Planning for the tricentennial began in 1991 and continued through to 1996.[86] Major commemorative events were held at Rottnest Island, the Swan River, and Dirk Hartog Island, together with a variety of other activities, all of which were well received by the public.

The Rottnest Island function was held on 29 December 1996 on the beach at Thomson Bay.[87] It was a beautiful day, and the function included an impressive and entertaining re-enactment of Willem de Vlamingh's landing. Peter Holland delivered a vivid reading of the history of Vlamingh's visit, and the Premier of Western Australia, Richard Court, gave a speech and unveiled a plaque to commemorate Vlamingh's landing at The Basin. The event was enjoyed by all, and its story will, I am sure, become part of the rich folklore of Rottnest Island's history.

The Swan River celebration was held on 12 January 1997. The event began with a voyage up the river by a flotilla of small boats, retracing Vlamingh's visit of 300 years before. Landings were made at Freshwater Bay and Matilda Bay, where crowds had gathered to witness the spectacle. The culmination of the occasion was a ceremony at Burswood Park, where the Prince of Orange, Willem-Alexander, the Crown Prince of the Netherlands, delivered a speech and unveiled a statue of Willem de Vlamingh. The large crowd included many people from the Dutch community in Perth, some dressed in traditional Dutch costumes. It was a memorable day, which will be remembered with pleasure by all who attended.[88]

The culmination of the tricentennial celebrations was held at Dirk Hartog Island on 12 February 1997. Planning for the event presented a big challenge to the organising subcommittee, which I chaired. We were well aware of the difficulties that might be encountered in holding an appropriate ceremony at this remote and inaccessible locality. Strong southerly winds prevail during much of February, and consequently, the transportation of large numbers of people to the site could be a logistical nightmare. Moreover, the weather could be excessively hot. However, we decided that it was worth taking the risk, as the function had the potential to constitute a spectacular conclusion to the Willem de Vlamingh Tricentennial.[89]

Fortunately, the weather on the day of the event proved to be relatively calm. Some of the 200 invited guests were flown directly to the site by helicopter, while others flew to Dirk Hartog homestead in a light aircraft and were driven overland by four-wheel-drive vehicle, travelling over a rough bush track for some 65 kilometres to reach Cape Inscription. Other people came by four-wheel-drive vehicles that had been transported to the island by barge a few days previously. Most guests arrived in a flotilla of boats, which had left Denham at daybreak to make the 75-kilometre crossing to the island. They were ferried ashore by landing barge, and then lifted from the beach to the lighthouse by helicopter. Some of the schoolchildren enjoyed the helicopter ride so much that they immediately scrambled down the cliff to repeat the performance!

On the day of the ceremony, HMAS *Moresby*, a Royal Australian Navy hydrographic survey ship, anchored off Cape Inscription, brightly decorated for the occasion. Officers of the ship, in white dress uniform, were flown to the site by a naval helicopter. For some time, there were three helicopters busily taking off and landing beside the lighthouse, making the scene more like that of a military operation than a civilian celebration! However, everything eventually fell into place, and the ceremony itself began on schedule.

After my opening address to outline the history of the locality, speeches were delivered by the Premier of Western Australia, Richard Court; the Ambassador of the Netherlands, Dr Roelofsz Smit; and the President of the Shire of Shark Bay, Les Moss. The speeches were followed by an impressive re-enactment ceremony beside the cleft in the rock where Willem de Vlamingh's post and plate had been placed 300 years before.[90] An appropriate cypress tree trunk had been obtained from Garden Island (as none are now available on Rottnest Island), and hewn into a suitable shape for use in the re-enactment ceremony. A bronze plaque to commemorate Vlamingh's visit and a replica of his pewter plate had also been prepared for the occasion.[91]

The Dirk Hartog Island event proved to be a fitting finale to the tricentennial celebrations, and I believe that all who participated were left with a much greater appreciation of the history of this fascinating place.

As for Willem de Vlamingh himself — could he ever have fancied that his expedition would become cause for celebration so long after the event? Although he may have expected that his voyage would become widely known in the Netherlands, I am sure he could never have imagined, even in his wildest dreams, that it would be commemorated by English-speaking people, 300 years later, in that far-away land of Terra Australis.

Notes

[1] The name 'Terra Australis' is generally used in this book for the southern continent, in accordance with the usage in Willem de Vlamingh's personal journal. Most Dutch East India Company documents at the time used the name 'Zuyd Landt', meaning 'South Land'.

[2] Soon after the expedition was despatched, rumours were received at the Cape of Good Hope that the *Ridderschap van Holland* had been taken by pirates based in Madagascar (Halls, 1965). However, it appears that these rumours were afterwards discounted, and the fate of the ship remains a mystery to the present day (Playford, 1996).

[3] The names 'the *Geelvinck*', 'the *Nyptangh*', and 'the *Weseltje*' are given in Dutch as '*De Geelvinck*', '*De Nyptangh*', and '*'t Weseltje*' (shortened form of '*Het Weseltje*'). Throughout this book, the names are given as *Geelvinck*, *Nyptangh*, and *Weseltje*.

[4] English-speaking authors who discuss Willem de Vlamingh and his voyage nearly always use 'Vlamingh' (or 'Vlaming' in older books) as the contraction of his name. On the other hand, 'De Vlamingh' or 'de Vlamingh' are both used by modern Dutch-speaking authors writing in English, and 'De Vlamingh' is now the accepted usage in the Dutch language. However, the criterion for correct practice in English is usage in spoken and written English, not usage in another language. Consequently, 'Vlamingh' is used as the appropriate contraction of Willem de Vlamingh's name throughout this book.

It is interesting to note that the first English-speaking author to use the contraction 'Vlaming' was Flinders (1814). In doing so, he was following the well-known eighteenth-century Dutch authors Witsen (1705) and Valentyn (1726), and the French author Prévost d'Exiles (1758), all three of whom used 'Vlaming', rather than 'De Vlaming'.

The name 'Vlamingh' has also been applied to all the geographic features named after Willem de Vlamingh in Western Australia, and it is notable that among the first and last features of the Australian coast seen by him, two now bear his name: Cape Vlamingh at the west end of Rottnest Island, and Vlamingh Head near North West Cape. Other features using Vlamingh's name include Vlamingh Memorial, Vlamingh Lookout, Vlamingh Way, and Vlamingh Cottage, Rottnest; Vlamingh Memorial, Cottesloe; Vlamingh Parade, Mosman; Vlaming Rise, Coogee; and Vlamingh Parklands, North Fremantle. The French also named Pointe Vlaming on Amsterdam Island in recognition of his role in exploring that island. Five memorials (or memorial plaques) have been erected in Western Australia in honour of Vlamingh, at Cottesloe, Burswood Park, Rottnest Island, and Dirk Hartog Island.

[5] The principal published sources of information on the voyage that were consulted during my study are Witsen (1705), Leupe (1868), Stapel (1937), Robert (1972), Sigmond & Zuiderbaan (1976a, 1976b), Schilder (1976a, 1976b, 1978, 1984, 1985), and Appleyard & Manford (1979). All these accounts are based primarily on the log of the *Nyptangh* (written by Mandrop Torst, and first published anonymously in 1701) and/or the log of the *Geelvinck*. Other early descriptions of the voyage by Valentyn (1726), Prévost d'Exiles (1758), and Flinders (1814), were apparently based on another journal of the expedition, which is now lost.

[6] The transcript obtained from France by the Parliamentary Library is now held in the National Library of Australia, Canberra, Reference No. NLA MS 11 Folder 20.

[7] Letter dated 5 May 1997 to Phillip Playford from Maïté Etchechoury, with an attachment by Phillipe Henrat, of the Archives de France. The original French manuscript in the Archives de France is registered as Marine 3 JJ 391 no. 18, and the document held by the Battye Library, Perth, is a partial transcript of that manuscript (covering Vlamingh's voyage along the coast of Terra Australis, but excluding the earlier part of the voyage). It had been transcribed in 1912 from the original in Paris by Madame Robert Hélouis, on behalf of the Commonwealth Parliamentary Library. Madame Hélouis hand-copied a total of some 6000 pages of early French journals relating to Australia—a remarkable achievement. All of her transcripts are now held in the National Library of Australia, Canberra.

I later obtained a microfilm copy of the original manuscript held in the Archives de France. The Marine Nationale, Service Historique, Vincennes, France, holds a contemporary copy of the original French translation, registered as Ms 216, and I also obtained a microfilm copy of this manuscript. Another early copy of the original translation is stored in the Archives de France (Marine 3 JJ 378 no. 24), marked as having been prepared for Monsieur d'Entrecasteaux, no doubt for his voyage of discovery along the Western Australian coast in 1792. A further manuscript (Marine 3 JJ 378 no. 25) is a partial copy, covering most of the voyage along the Western Australian coast.

In the opinion of Phillipe Henrat of the Archives de France, the original French translation of Vlamingh's journal may have been prepared during the period 1730–40. On the other hand, Andrew Hunwick of Perth, who is familiar with French manuscripts of the eighteenth century, thinks that the translation might be older—dating perhaps from the first decade of that century.

A barely legible handwritten statement at the front of the Marine Nationale copy has been translated by Andrew Hunwick as: 'Contemporary copy. A slightly more recent hand has entered on the first page: Journal of the voyage of W. Vlaming'.

[8] See Stapel (1937) and Somerville (1948).

[9] See Robert (1972), Sigmond & Zuiderbaan (1976a, 1976b), Schilder (1976a, 1976b, 1984, 1985), and Appleyard & Manford (1979).

Schilder made reference to the existence of the French manuscript (1985, Ch. 5, note 4, pp. 236–7), but he stated, incorrectly, that it is a translation of parts of the *Geelvinck* log and made no further comment on it.

Stapel speculated that Nicolaes Witsen had loaned Vlamingh's journal to one of his colleagues, who translated it into French. However, Witsen's publication (Witsen, 1705, translated in Schilder, 1985), gives no indication that he ever saw the journal. Perhaps it remained among Vlamingh's personal possessions when he died, and eventually passed into French hands after the sale of those possessions. The original manuscript in Dutch must have been lost after it was translated into French.

[10] I was unaware of the existence of this French manuscript when I presented my lectures on Willem de Vlamingh during the tricentennial. Those lectures were given at the Western Australian Maritime Museum, Fremantle, on 16 June 1996 (as part of a series entitled 'Dutch Discoveries and Shipwrecks, 1616–1727'); the Royal Society of Western Australia on 13 December 1996; Rottnest Island on 28 December 1996; the Western Australian Maritime Museum on 13 January 1997 (as part of the Australian–Dutch colloquium on maritime history and archaeology—see Playford, 1998); and Denham on 9 February 1997.

[11] The first Europeans known to have set eyes on the Australian continent were men on the Dutch ship *Duyfken*, who charted the west side of Cape York Peninsula in the Gulf of Carpentaria in 1606 (Schilder, 1976b). Subsequent accounts of that voyage reported that some men on the *Duyfken* also landed there (Heeres, 1899), but this cannot be proved as the ship's log is lost. Moreover, it is clear that the skipper of the *Duyfken*, Willem Jansz, believed that all the land he had encountered formed part of New Guinea. Consequently, his discovery had no significant impact on world cartography.

On the other hand, Dirk Hartog's discovery in the *Eendracht* of the west coast of Australia in 1616 resulted in Terra Australis being shown factually on world maps for the first time, commonly as the 'Land of the Eendracht', and later as 'The South Land' and 'New Holland'. With regard to Hartog's discovery, Schilder (1976b) commented that 'Hartog's ship's log is lost, but there are a number of contemporary letters in the archives of the VOC which confirm his discovery. The absence of direct documentary evidence from the ship itself makes all the more important the one surviving material evidence of the discoveries made by the *Eendracht* – a pewter plate of 1616 which Hartog fastened on a pole to commemorate his landing on the island'. Thus, in the case of Dirk Hartog's visit there is material proof, in the form of the famous pewter plate, that a landing was made. However, there is no unequivocal evidence that men of the *Duyfken* landed in Australia ten years before, during their voyage in the Gulf of Carpentaria (see *West Australian*, 'Hartog should get credit for first landing', 19 Feb. 1997).

[12] See Flinders (1814).

[13] See Blainey (1995).

[14] See Schilder (1985).

[15] See Schilder (1985).

[16] See Appendix 4 in Schilder (1985): 'Surgeons' journals of the *Geelvinck* and *Nyptangh*'.

[17] See Jacobs (1991).

[18] For a graphic description of the effects of scurvy and the rapid recovery effected after eating fresh vegetables, see Dana (1840).

[19] Letter from Nicolaes Witsen to Gisbert Cuper (1698), reproduced as Appendix 13 in Schilder (1985).

[20] See Schilder (1985).

[21] Witsen's statement that he hoped that the expedition would manage 'to catch a South Lander and bring him hither' was contained in a letter written by Witsen to Gisbert Cuper in 1696 (Schilder, 1985).

[22] See Heeres (1899) and Loos (1994).

[23] See Playford (1996).

[24] See Stapel (1937).

[25] This map is shown on page 1 of volume 15 of the book series on world voyages compiled by Prévost d'Exiles (1758). Volume 15 deals with early exploration of Terra Australis. See also Schilder (1985).

[26] See Schilder (1985).

[27] See Heeres (1899).

[28] See Schilder (1978, 1984).

[29] Letter from the Governor-General and Councillors of the Indies to the Heeren XVII, 30 November 1687, KA VOC 1587, pp. 402r–407v, quoted in Schilder (1985).

[30] See Somerville (1948) and Seddon (1972).

[31] See Appleyard & Manford (1979).

[32] Published in Schilder (1978, 1984) and Appleyard & Manford (1979).

[33] The principal difference between the north side of Rottnest Island as it can be seen today and as it was depicted in Victorszoon's sketch, is that most of the forest cover that was present in 1696 has now disappeared. Also, Phillip Rock has much less relief today—the top of the rock was cut off during World War II so that it would not obstruct searchlights mounted at Phillip Point.

[34] This information was reported in the *Rottnest Islander*, no. 62, Dec.–Jan. 1996, pp. 8–9. A plaque identifying The Basin as Vlamingh's landing point was unveiled by the Premier of Western Australia, Richard Court, on 29 December 1996.

[35] See maps in Schilder (1984).

[36] See Witsen (1705), translated in Schilder (1985).

[37] See Henderson (1982).

[38] For a discussion of the freshwater resources of Rottnest Island, see Playford (1988, 1997). It seems clear that *Salmonella* bacteria, which now infect the local fauna and contaminate the untreated groundwater at Rottnest Island, were not present in 1696, otherwise Vlamingh's men would have become ill after drinking the water.

[39] There were 15 Dutch miles to a degree of latitude. Thus, a Dutch mile was equivalent to 4 nautical miles of modern usage, or 7.408 kilometres.

[40] See King (1827).

[41] Prior to European settlement, the cypress forest and most of the other vegetation covering Rottnest Island was periodically destroyed by major bushfires, which were ignited by lightning. Most of the quokkas on the island perished in such fires, and consequently, the dense cypress seedlings that sprang up after each fire were not subjected to extensive animal browsing. By the time a large quokka population was re-established, some years after the fire, the cypress trees in the new forest would have grown large enough to survive browsing. As a result of this regime of periodic major fires followed by forest regeneration, almost all cypress trees on the island would have once been the same age, with approximately the same dimensions, thus giving rise to the 'monotonous' appearance reported by Cunningham (King, 1827).

The same process occurred on Garden Island after a major fire during the 1950s which killed most of the tamar wallabies in the area. This allowed a dense cypress forest to grow up in the following years. Thus, the type of cypress forest that was once widespread on Rottnest Island can still be seen on Garden Island.

[42] See Schilder (1985).

[43] Water depths were measured by the Dutch in fathoms of 6 Amsterdam feet (1.698 metres).

[44] Otherwise *Acacia rostellifera* or *Acacia cyclops* (R. Powell & N. Marchant, personal communication, 1997).

[45] General report from the Governor-General in Batavia dated 2 February 1698, KA VOC 1589, p. 1622, quoted in Schilder (1985).

[46] See Sigmond & Zuiderbaan (1976b) and Appleyard & Manford (1979). The Vlamingh Memorial, constructed by the Town of Cottesloe, was opened by the Governor-General of Australia, Sir Paul Hasluck, on 5 January 1974. Unfortunately, the original construction materials corroded badly in the exposed coastal position, and in 1997 the main structure was demolished, leaving only a column bearing the name 'Vlamingh' and the original plaque (see the *Post*, 'Vlamingh Memorial plan scrapped', 31 May 1997). Two new plaques were unveiled at the site by the Minister for Planning, Graham Kierath, on 16 November 1997, to commemorate the 300th anniversary of Vlamingh's visit and to mark the dedication of the Vlamingh Parklands.

[47] Lake Claremont was known as *Katabberup* to the Nyungar Aborigines and Butler's Swamp to the early European settlers.

[48] See Valentyn (1726) and Flinders (1814).

[49] Hammond (1933) described the way in which tall trees were climbed by Aborigines to catch possums, using notches cut into the trunks with stone axes. When climbing a tree in this way, an Aborigine held an axe in each hand, using them alternately to cut steps for his toes. When preparing to chop a new foothold with one axe, the pointed handle of the other was driven into the bark of the tree, to hold the climber steady.

[50] I am indebted to George Seddon for discussions regarding the climatic implications of Vlamingh's voyage and their relationship to the Little Ice Age.

[51] See Witsen (1705), translated in Schilder (1985).

[52] See Valentyn (1726).

[53] The Nyungar Aborigines of the south-west had learned to detoxify these nuts, known as *by-yu*, by prolonged soaking in water (see Grey, 1841 and Hammond, 1933).

[54] The caption used by Valentyn (1726) to describe the sketch of the three vessels shown in Figure 13 is '*Swarte Swaane drift op het Eyland Rottenest*', which can be translated as 'Black swan[s] floating on [or near] the Island Rottenest'. However, the scene is clearly supposed to be the mouth of the Swan River. Note that the *Geelvinck* and *Nyptangh* are depicted in this sketch with topgallant masts, which are lacking on the drawings of those ships on Victor Victorszoon's map (see cover illustration and Figure 6). The topgallant masts must have been lowered when the ships were sailing over the Indian Ocean before strong westerly winds, and were raised after the ships reached Rottnest Island, where the winds were less violent. The *Geelvinck* is also shown in the Valentyn sketch with its gunports open along the side, whereas those ports cannot be seen on the other drawing. This is because the side gunports were normally closed when sailing on the open ocean, in order to keep the crew's living quarters dry, but were opened while lying at anchor, to facilitate ventilation.

Another version of this Valentyn sketch was published by Prévost d'Exiles (1758). That version showed only the *Geelvinck* and the *Weseltje*, omitting the *Nyptangh*, and the French translation of the caption was given as '*Canal aux Cygnes Noirs dan l'Isle Rottenest*'.

[55] Either a reed warbler, rufus whistler, shrike thrush, or willy wagtail according to S.J. Davies & K. Coate (personal communication, 1997).

[56] See Witsen (1705), translated in Schilder (1985).

[57] I am indebted to George Seddon for this suggestion.

[58] See Bates (1992) and Green (1984).

[59] See Witsen (1705), translated in Schilder (1985).

[60] See Witsen (1705), translated in Schilder (1985). Although Witsen must have acquired Vlamingh's original maps, they have since been lost, so Witsen's claim regarding the naming of the 'Witsen River' cannot be confirmed. That name was not used in Vlamingh's journal or either of the ships' logs.

[61] Aboriginal fire practices were a source of conflict with the early British settlers in Western Australia. The lighting of bushfires was strongly opposed by the settlers, especially during summer, as it destroyed their pasture, crops, and fences, and endangered their homes (see Moore, 1884 and Lines, 1994).

The use of fire in bush and forest management is still a controversial issue in Western Australia. Whereas prior to British settlement the natural vegetation had evolved in a regime of regular summer fires, no burning is now permitted at that time of the year, because it poses a serious threat to life and property. This change in the burning regime is leading to significant changes in the flora and fauna of the native bush, which will continue until a new equilibrium is eventually achieved.

[62] I am indebted to Keith Hockey, Kieren O'Connor and John Erickson for advice and assistance in relation to a visit that I made to examine this area.

[63] Parts of Hutt Lagoon are still often red in colour, due to the blooming of microscopic algae of the species *Dunaliella salina.* These algae are now cultivated in specially constructed ponds beside the south-eastern margin of the lagoon, and are processed commercially as a source of beta-carotene (used as a colouring agent in foods). Prior to this activity, the lagoon used to dry out almost completely during the summer months, becoming white through the precipitation of salt on its bed. However, sea water is now pumped into the lagoon to maintain the beta-carotene plant, so that the lagoon no longer dries out fully during the summer months, and parts of it remain red throughout the year.

The salt on the floor of the lagoon was mined commercially during the early 1900s. It is therefore interesting to note that two of Vlamingh's observations at Hutt Lagoon related to products that would eventually be developed commercially—salt and beta-carotene.

[64] See Witsen (1705), translated in Schilder (1985).

[65] See Playford (1996).

[66] Wittecarra Spring in Kalbarri National Park is among the most historic localities in Western Australia. During a visit there in September 1996, on an excursion for the Willem de Vlamingh Tricentennial, I was very surprised to find the spring dry. On previous visits it had always contained a pool of fresh water, even during the hottest part of the summer. I learned afterwards that shortly before our visit, the Main Roads Department had extracted large volumes of groundwater from a nearby bore, for use in road-building operations. This had lowered the regional water table, causing the spring and other water seepages along Wittecarra Gully to become dry. Large-scale extraction of water from the bore has now ceased, but several years of normal rainfall may be needed to reactivate the spring.

[67] George Grey is the first European known to have landed in Gantheaume Bay after Vlamingh's visit (Grey, 1841).

[68] See Playford (1996).

[69] It is surprising that a facsimile of the inscription on the Hartog plate has not previously been published. The interpreted facsimile shown on Figure 18 was made using photographs and x-ray images of the original plate, supplied by the Rijksmuseum, Amsterdam, and in consultation with Dr J.P. Sigmond of that museum. However, a number of uncertainties remain. The wording is in accordance with Freycinet's transcription of the relevant part of Vlamingh's plate (Freycinet, 1825), except that I have used 'Pieter Dookes van Bill' rather than 'Pieter Doores van Bil' of Freycinet's transcription. Note that Balbian Verster (1899), in his interpretation of the Hartog plate, uses 'Dockes' and 'Dooke(?)' instead of Freycinet's 'Doores'. Usages by other authors are 'Dookus' (Valentyn, 1726), 'Dookes' and 'Dockes' (Robert, 1972), and 'Dockes', 'Doekes', and 'Dookes' (Schilder, 1985). The spelling 'Bill' seems to be the best interpretation of the relevant part of the very faint scratched lettering remaining on the original Hartog plate, and that usage is adopted here. It is also the usage of Valentyn (1726) and Robert (1972), although other authors have generally used 'Bil'. There is no town named either 'Bill' or 'Bil' in the Netherlands ('Bil' means 'buttocks' in Dutch), but there is a River Bille near Hamburg, not far from the Netherlands, and the uppersteersman Pieter Dookes may have come from there. The name of this person must have been very lightly scratched on the pewter plate, and even in 1697 it may have been only weakly discernible. Today, it has almost totally disappeared. Thus, it is impossible be sure of the correct usage, but I have adopted 'Pieter Dookes van Bill'.

Balbian Verster (1899) omitted 'Anno 1616' from the end of his interpretation of the Hartog plate, contrary to the text shown on the Vlamingh plate. However, the 'A' of 'Anno' is still visible on the original plate, and therefore I have included 'Anno 1616' at the end of my transcription. Valentyn (1726) used 'Ao. 1616', so that is another alternative.

Note that Dirk Hartog spelled his name as 'Dirck Hatichs' on the original plate, whereas some VOC personnel in the early seventeenth century wrote it as 'Dirck Hartochsz' and Vlamingh used 'Dirck Hartoogs'. Subsequent Dutch authorities usually referred to him as 'Dirk (or 'Dirck') Hartogs', 'Dirk Hartogsz', or 'Dirk Hartogs(zoon)' (for example, Heeres, 1899). However, the usage 'Dirk Hartog' is now generally accepted in English.

[70] See Balbian Verster (1899).

[71] Note that '12th ditto' is not included in this way on the original inscription, and that is why it is shown in brackets in the translation. However, it is clear that '12th ditto', i.e. the 12th of February, is the meaning that was intended for the inscription '#12' shown at the base of the original plate (see Figure 19).

[72] In 1699, two years after Vlamingh's visit, William Dampier, the renowned British adventurer and scientist, anchored off the north-eastern shore of Dirk Hartog Island, probably between Cape Levillian and Withnell Bay. He named the great bay where he was anchored 'Shark's Bay', because of the many sharks that his men caught and could see in the water. For details, see Marchant (1988).

Dampier's plant collection from the island is the first scientific collection to have been made anywhere in Australia, and it is still preserved in the collection of Oxford University.

It is hoped that the collection will be returned on loan to Australia in 1999, for display during the tricentennial of Dampier's visit to Shark Bay.

[73] Another French explorer, François de St Allouarn, visited Dirk Hartog Island in 1772. He landed at Turtle Bay, about 2 kilometres from Cape Inscription, but did not seek to locate the Vlamingh plate and post. However, St Allouarn's men buried a bottle at the foot of a tree beside the cliff top near where they landed, containing a parchment with a written statement claiming the western side of New Holland in the name of the King of France. Two silver coins were buried with the bottle. For details, see Marchant (1982) and Horner (1987).

On 20 January 1998, it was reported that a French silver écu coin, dated 1766, and wrapped in an ancient lead bottle cap, had been found buried below a tree beside the cliff top above Turtle Bay. This coin is almost certainly one of the two left there by St Allouarn (see *West Australian*, 'French coin may solve history riddle', 20 Jan. 1998, p. 3). On 2 April 1998, a further report indicated that a sealed bottle had been uncovered nearby, following a systematic survey using metal detectors, and that it may prove to contain the remains of St Allouarn's proclamation (see *West Australian*, 'Bottle found at coin site', 2 April 1998, p. 7).

[74] See Freycinet (1825).

[75] The first of such requests was initiated by the Commonwealth Parliamentary Library Committee in 1912. The minutes of the Committee for 21 August 1912 recorded that 'Correspondence relating to a suggestion of Sir Arthur Morgan that the French Government might be asked to present to the Commonwealth the brass [sic] plate erected by Vlaming in 1697 at Dirk Hartog's Island and removed thence to Paris by Freycinet was laid before the Committee and on the motion of Mr Groom it was decided to send a letter to the Prime Minister asking him to open negotiations with a view to obtaining possession of the plate mentioned.'

[76] See Renié (1944).

[77] See King (1827) and Hordern (1997).

[78] Memorandum dated 18 June 1907 to the Engineer in Chief, Perth, from W.H. Shields, Resident Engineer. See Public Works Department file 4032/1907, copied in Royal Western Australian Historical Society file HS/540.

[79] See Royal Western Australian Historical Society file HS/540, Dirk Hartog Island.

The concern regarding cliff stability was first raised by geologist Arthur Wade in 1936, after a visit to the site. He suggested that the crack in the rock where Hartog, Vlamingh, and Hamelin erected their posts had been widened appreciably when the two jarrah posts were driven into it. Consequently he thought that the site was threatened by imminent collapse. However, I doubt that there was any sound basis for that conclusion. The crack was already open when Dirk Hartog visited there in 1616 (that was why he was able to put his post in a 'cleft in the rock'), and there is no reliable evidence that it has expanded substantially since then, or that the cliff face is likely to fall in the near future. Although natural agents of erosion will continue to wear away the cliffs, and the segment of the cliff adjoining the crack will eventually collapse, this may not occur for many years. However, for public-safety reasons the width of the crack should be monitored at regular intervals, and it will be important to ensure that erosion is not accelerated by human activities.

[80] The suggestion for such a memorial was first made by Mr H.J. (Joe) Spaven, then Chairman of the Shark Bay Road Board, in a letter to the Historical Society dated 10 April 1937 (Royal Western Australian Historical Society file HS/540). He considered that 'As this locality is the most historic spot of Western Australia [and] even of Australia, a suitable memorial should be erected ... to which I will contribute £5'. He stated that the site should 'be made a place worthy of pilgrimage for tourists', and that it could be arranged for the Roads Board to construct the necessary access road to the site. Spaven went on to observe that 'If novelists like Zane Grey want to write a historic story, where could they find more for a story of romantic adventure? The names on the chart are sufficient. French, Dutch, and British adventurers have all been here [and] even a few Australians, ... buccaneers, like myself'. He added that he, and others like him, were only referred to as a buccaneers 'by nasty-minded people who do not understand our high-minded motives, which are to get the necessary [funds] to buy food and more beer'.

The Memorials Committee of the Historical Society resolved that publicity should be given to Mr Spaven's proposals, and representations were made for funding from the State and Commonwealth Governments. The Commonwealth eventually agreed to cover the full costs (£50) of an appropriate plaque to be made and installed on the Cape Inscription Lighthouse.

The wording to be used on the plaque was decided on by Dr J.S. Battye in close consultation with a leading Dutch archivist, Dr F.R.J. Verhoeven. It included the statement that Dirk Hartog's landing was 'the first known landing of Europeans in Australia'.

The lighthouse keepers' quarters, currently lying in ruins, were built in 1910, but were abandoned only seven years later when control of lighthouses in Western Australia was passed to the Commonwealth Government and the light was automated. However, the concrete walls are still sound, and a proposal for restoration of the quarters was first put forward in 1992 by Les Moss and Leon Green (*West Australian* 9 November 1992). A Conservation plan for possible future restoration of the quarters was completed for the National Trust in 1996 (when I was the Trust President), with funding from the Lotteries Commission. I believe that if the quarters can be properly restored and an airstrip is installed nearby, this historic site could become one of the premier tourist attractions in Western Australia. However, it will be essential to ensure that if the project proceeds, every effort is made to ensure that the historical, ecological, and scenic values of this remarkable site are adequately protected.

[81] See Schilder (1985).

[82] Letter from Nicolaes Witsen to Gisbert Cuper (1698), quoted in Schilder (1985).

[83] Letter from Nicolaes Witsen to Martin Lister, 3 October 1698, quoted in Schilder (1985).

[84] See Witsen (1705), translated in Schilder (1985).

[85] The prominent Australian historian, Geoffrey Blainey, has listed the discovery of the Swan River in 1697 as one of the principal events in Australian history—the only Dutch discovery that he recognises in this way (Blainey, 1995).

[86] The proposal for celebrations to be held to commemorate the Willem de Vlamingh tricentennial was first made at a meeting of the Rottnest Island Environment and Research Committee on 26 June 1991. An organising committee was then set up to plan the tricentennial, with myself as chairman. Many of the functions that finally eventuated were suggested at the inaugural meeting of the committee held on 18 September 1991. Subsequently, because of other commitments that required me to be absent from Perth for long periods, I elected to stand down as chairman, and that role was passed to Thom Dercksen.

[87] Brian Easton chaired the Rottnest organising committee and Merri Blakemore planned and directed the ceremony.

[88] The Swan River celebrations were directed by Mara Basanovic of Celebrate WA (a State Government authority), who was also a member of the tricentennial committee.

The tricentennial celebrations were not without public controversy, involving supposed Government cost-cutting restrictions (see *West Australian*, 'Cost cuts hurt Swan event', 11 Jan. 1997). However, in my opinion, the financial and other assistance was adequate, and the Western Australian State Government, local governments, and several governmental authorities played major roles in the success of the tricentennial events.

[89] Secretarial and other assistance was provided to the organising subcommittee by the National Trust. The Shire of Shark Bay accepted responsibility for the logistical arrangements for the event through the Shire President, Les Moss, who was also a member of the organising subcommittee. The Shire arranged for some 200 invited guests (including many schoolchildren) to be transported to and from the island, and to be provided with shelter, lunch, and drinks. The Shire staff performed this demanding task very effectively. Funding for the event was provided by the State Government, the Shire of Shark Bay, and the Lotteries Commission.

[90] The re-enactment was directed and produced by Dina Garrett and Mike McCarthy of the Western Australian Maritime Museum, Fremantle, with the assistance of other people from the Museum and the town of Denham.

[91] The bronze plaque and replicas of the Willem de Vlamingh and Dirk Hartog plates were engraved by Poolegrave Engravers, using my designs (Figures 18, 19, and 21). The pewter plates were cast by Christo Pewter, and funding was provided by the Lotteries Commission. The replica plates have since been placed in the collection of the Western Australian Museum.

APPENDIX

The Journal of Willem de Vlamingh
(Translation from the French by C. de Heer and C.I. Clerk)

Willem de Vlamingh's journal of 1696–97 is known only from a handwritten French manuscript, translated from the Dutch during the early eighteenth century. The original Dutch journal is thought to be lost. The French translation is held in the Archives Nationales de France, as document Marine 3 JJ 391 no. 18. Another contemporary copy is held in the Marine Nationale, Service Historique, Vincennes, as document Ms 216. These manuscripts are entitled: 'Observations et découvertes faites pendant les années 1696 et 1697 vers les Isles de Tristan da Coha, St. Paulo, et Amsterdam, et le long des côtes de la terre Australe ou inconnüe' ('Observations and discoveries made during the years 1696 and 1697 about the islands of Tristan da Cunha, Saint Paul, and Amsterdam, and along the coast of the Terra Australis or unknown').

The English translation that follows covers only the section of the journal dealing with exploration of the Western Australian coast, on pages 42 to 118 of the Archives Nationales document.

Account of our arrival at the Terra Australis, still unknown, at the end of the year 1696 and the beginning of the year 1697

On the 23rd [sic, should be 28th] the wind south-east, with a topgallant breeze, and at noon our estimated course and distance in 23 days had been east ¼ north-east, and this, having brought us to latitude 31°19' south and longitude 131°46', we found by observations that the current was moving south with a calm sea, and we had a variable wind from the south ¼ south-east, south ¼ south-west and south-south-west, with a topgallant breeze. Here we spotted a quite extraordinary bird called a gull, which had a black head and wings, the rest of the body being completely white, and from this judged that we were not far from the coast. This is why we cast the lead quite often after mounting the first watch and found depths of 65 fathoms with white sand. This gave us an extraordinary joy, the more so as the great God had favoured us with unparalleled care, and, to glorify him, we allowed ourselves to drift for a period of four hours, the wind south-south-west with a slight strength, and the depth as before except that the bottom was a form of chalk. We set the sails according to the wind and travelled thus with little sail up to the day-time watch, the wind being as before. We cast the lead at 35 fathoms and also from 40 fathoms to 30 fathoms where there was a sea-bed of red stones like coral.

The 29 December at midday, the wind north-west, light breeze, we set course south-east and had a depth of 30 fathoms with a sandy bottom. We saw floating tree branches as well as tree stumps.

At midday our estimated course and distance in 24 hours was east ¼ south-east, nine miles, and this having brought us to latitude south 31°48' and longitude 132°28'. We found by observation that we were in fact at 31°47'. After running for the duration of another hourglass we saw the Isle of Girls ahead of us, and at once steered for it, and taking soundings we observed that the depth diminished suddenly from 30 fathoms to only 10 fathoms where we dropped anchor through God's special blessing.

On Sunday [sic, actually Saturday] evening, after sending our boat to sound whether there was danger around there from any sharp or rocky bottom, and finding it everywhere to be very good, we were nevertheless forced to break the resolution which we had made to go ashore and to amuse ourselves with exploring the island and admiring the fine trees which we saw in the distance, as the darkness of the night robbed us of this pleasure.

After passing the night, in the morning of the 30th, before daybreak, the wind from the south, with a topgallant breeze, not too full. As soon as the morning dawned, we deemed amongst ourselves, because we must not lose time, that we ought to send out one of our boats to row all around the island to see if it could not discover any remains of shipwrecks or debris of ships, and to despatch another with the bookkeeper, the uppersteersman, and the commander of the soldiers and twelve men, with orders to go on land and to stay there until nightfall to carry out an exact reconnaissance of the said island.

When all had received their orders, they set out. The last boat returned on board in the evening with the report that the bookkeeper and the other people had explored much of the island, but they had observed there only trees, and having realised that they had already covered quite a distance on shore without discovering anything else, they had not deemed it proper to carry on further, and as they were heading back they had spread out from their path to go to a thicket of trees. Having arrived there they found that the leaves of the trees had a very pleasing sweetness. They cut some lengths of wood, and after killing some forest rats, they once again took to their path.

Having at last arrived at a very low-lying plain, and being oppressed by thirst, they had scooped out some little holes, from which there came very pleasant water to drink. But the arrival of night had surprised them, and half an hour after sunset they could discover nothing more. The people in the other boat returned on board after rowing all around the island without having found or noticed any ships' debris.

On the 31st, having heard the favourable report of the said island, I was urged by my people to go there in order to carry out a more detailed examination. With this in mind, and to satisfy my curiosity, I went ashore and landed without any trouble. Then having given my orders to the men and those who were to explore the island, my bookkeeper and I set off on our way and after marching about one hour we noticed a very attractive and thick grove of trees of about half a mile in extent. Its trees were arranged in such a way that one would think they had been planted in a straight line by a professional, to make it into a park. From these trees we could smell a very pleasant fragrance, which was spread all over the island.

We would have liked to press on further, but feeling tired already from the short walk I ordered the bookkeeper to proceed on his own and to explore more widely what the island had to offer. At the same time I resolved, together with some other people who were with me, to rejoin our wood-cutters.

On my way back I had great pleasure in admiring this island, which is very attractive, and where it seems to me that nature has denied nothing to make it pleasurable beyond all islands I have ever seen, being very well provided for man's well-being, with timber, stone, and lime for building him houses, only lacking ploughmen to till these fine plains. There is plentiful salt, and the coast is full of fish. Birds make themselves heard with pleasant song in these scented groves. So I believe that of so many people who seek to make themselves happy, there are many who would scorn the fortunes of our country for the choice of this one here, which would seem a paradise on earth.

While walking and strolling, I at last reached our woodcutters, and having rested some time with them, our bookkeeper arrived with a report that on the coast opposite us he had seen smoke, but that he had seen nothing else of importance. We all went back aboard together, with a slight breeze from the south-west, at the end of the day.

1697

From the 1st of January until the 3rd we sent men ashore every day to cut wood and collect fresh herbs. We were often brought some which were unknown to us, but which we found excellent. They also brought on board myrrh, aloes, and incense, and told us that they always saw constant smoke on the mainland, whereupon we decided to arm ourselves well to go there, and to explore everything closely.

On the 4th in the morning we sent our bookkeeper and our uppersteersman ashore to sound the mainland coast, who returned in the evening reporting to us that in a cove they had seen several small islands, 4 to 5 fathoms deep all around them. They had found sometimes a foul and bad bottom, and sometimes a fine and good one, and after much trouble they had finally come ashore, where they had found many trees dripping gum, of which they brought some aboard. And since they deemed themselves to be too weak in manpower, they had not dared to risk going any further ashore.

In the evening we held a meeting and it was resolved to detail the best 80 men from the three ships, officers, soldiers, and seamen, and to charge the officers in command of the party with the care of the whole enterprise, having appointed as chief and commander over all these people, Skipper Colaert, commander of the hooker, and his adjutants the bookkeeper Joannes Bremer and the uppersteersman Michiel Bloem; also from the said hooker Martinus Stypen, sergeant, as well as understeersman Jan Palensteyn and Claas Gerard, with their subordinate officers. They were all to go aboard the said galliot at daybreak the next day in order to be transported as close to the mainland as would be possible, and all were then to be put ashore by the boatmen.

For this purpose they were ordered to draw rations for a few days. I had planned to be present at this landing, but since we had no knowledge or description of the places and their location I deemed it proper to postpone going there until I could be informed of the success of this first landing before making another with more ease and providing the necessaries for the same.

On the 5th at daybreak our detailed party on the word of command went aboard the galliot and soon after it weighed anchor and set sail with a south-westerly wind, which as far as we could see took them to the shore.

On the 6th wind and weather as before, in the evening the wind west although in the morning it had been north-west.

On the 7th the wind north-north-west and north-west, topsail breeze, hollow sea. At midday our bower anchor parted; while making sail we observed the galliot with the boat which came to join us. Having moved forward close to a large shoal we came back through a depth of 16 fathoms and a grey sandy bottom to drop anchor. At night the wind north-west, topsail breeze, hollow and rough sea.

At noon the galliot came and dropped anchor close to us and we sent to take our men back from on board the galliot.

On the 8th the people who had been detailed to go ashore came on board and reported to me as follows: Having arrived ashore on the 5th by noon and an inspection having been held, they were found to number 86 men and the commander read out the orders of the Company signed by the General Council, and next having urged them to comply with the tenor of the orders, and [stating] that in case of a lack of this they could only expect severe punishments, they promised to carry them out in every possible detail and in accordance with their duty.

So they started off on their march, in the name of God, taking their course to the east on land, and after marching for an hour they came to a hill, where having seen a river in a glen they went down to it.

Having reached it they noticed footsteps of human beings around about and several stakes which must have been fixed in the ground. After half an hour, having gone forward along the stream, they came to three dilapidated huts which were like those of the Hottentots at the Cape of Maaksel, and not far from there they found several footprints of men in the sand, also of people of varying age and sex, such as girls, children and men, young and old, and of ordinary stature.

They continued their walk along the stream, but since they were growing tired and felt a need for rest and food, they made their way to an elevated place by the side of the stream and there made their rendezvous under a big, cool, green tree at least two fathoms wide, which was full of knots from top to bottom and was very easy to climb to the top. After resting and refreshing themselves, they marched off again all together and having covered some distance they saw three more huts. They returned from there to the bank of the river, alongside which they found several holes in the ground in which there was very good fresh water. They assumed that these had been made by the inhabitants, since the river water was a little brackish. Since, in their opinion, it was impossible to get back on board that day, they believed that they had to remain ashore all together, and at the same time they resolved to spend the whole of the next day on the exploration of the said land. So, after marching on about another half hour they went back to the three ruinous huts.

From there they observed smoke in many places straight ahead, and after talking it over they deemed it proper not to frighten the inhabitants by their large number but to send ahead 4 or 5 men and to follow them closely on one side.

As a result, the bookkeeper and the uppersteersman went forward with only two Indians, in the hope of meeting those inhabitants. But they found no-one, even though they realised from the footprints they had seen that inhabitants were undoubtedly in the neighbourhood, and this conviction had been confirmed from the huts they had found and the smoke seen rising up. From this it can be concluded that those people, on noticing ours, had withdrawn and hidden in the forests and barracks [sic]. However that may be, our efforts for this search have been useless up to now.

So, having wandered around for another three hours, they resolved to look out for a campsite which was convenient and secure. Having come to a very suitable place they set two sentries to prevent any surprises, and after giving instructions for hourly relief, they had their meal and rested.

The next day, at daybreak, they set forth once more, having divided their force into three sections or detachments, with orders to traverse thoroughly around the place of the rendezvous and to return there at a set time, hoping in this way to catch one of the inhabitants. At last the officer in command together with all the others marched forth, and after advancing for half an hour they came back again to the bank of the aforesaid river, which in this area seemed to them to be nearly three miles wide and also contained brackish but very clear water. Once again they saw several huts in the neighbourhood, but they admitted that it had been impossible for them to meet up with any inhabitants. However, they had seen various kinds of birds, white, green, and grey, of which they had killed some cockatoos and parrots with great curved beaks.

They went back from there to their meeting place, where upon their arrival they found the other two commanders who were cooking various fruits which they had gathered, and which made them sick after eating them.

So, having stayed there for some time for refreshment and rest, they started on their march again and divided their force in two, so that one of their party could pass on the inside of the river bank, whilst the other would follow on the outside. When they came to the edge of the bank they found there a piece of a ship's doubling, which was very old, and after walking for half an hour they came to where the sloop and the boat were at the mouth of the said river. They boarded the sloop and ordered those who stood watch over the boat that if contrary to their expectations the other party should arrive, they were at once to fire a few shots to warn them of their arrival so that they might make their way with them in the boat and take them there in the sloop.

After passing 4 or 5 shoals they found the river very deep and full of rocks and very difficult to enter. It was full of various kinds of birds, among which there were a number of young swans, all black, of which they had caught two, and there were also an abundance of fish of various species. On their return they sighted their comrades, and set their course straight for them. When they had joined them they told them that they had observed nothing of any importance, and since night began to fall they deemed it too late to go back on board. That is why they decided to remain still ashore this night and to return aboard at daybreak.

When dawn began to break they went aboard the galliot all together, and set sail with the wind north-west and west-north-west; they tacked like this until nightfall, when, as I have already reported, we all came to anchor together.

On the 9th of January, in the morning, the wind being south-south-west, we weighed anchor, set sail, and went to get the anchor we left behind as mentioned. We pulled it up with much trouble from between the rocks where it had become stuck, then we set sail once again, and set a course together between the island and the mainland. We came to anchor at 10 o'clock in the evening at a depth of 12 fathoms on a bottom of fine sand, and had wind and weather as before.

On the 10th of January, in the morning, the wind south with very pleasant weather. In the afternoon the wind coming from the sea, the galliot and the hooker came and dropped anchor at the same time close to us. I at once bade their senior officers to come aboard to take steps together in order to make a complete exploration of the said river. Therefore, having decided to be one of the party, I embarked to be taken ashore with the officers mentioned as well as soldiers and seamen detailed for this undertaking, as also with my bookkeeper, each having orders to go to his ship to draw the rations needed and then for all to go aboard the galliot which was to carry us as close to shore as possible, having with me then the uppersteersman of the hooker, my understeersman, and the third mate of the galliot on which all had arrived. We weighed anchor and ran ashore, assisted by soundings of 5 to 7 fathoms' depth, until we dropped anchor on a fine sandy bottom at 5 fathoms' depth.

Then we put 20 well-armed men on each boat, fitted with two swivels. I also embarked on a boat from the galliot, and so we came in a short time within the estuary, where we entered with much trouble because of the rocks there. We rowed together 5 to 6 miles up the river until 2 o'clock at night, and as our men felt tired we deemed it necessary to take a rest and to drop our grapnels so as to be more able to go and explore on land the next day, having made 5 to 6 miles.

The next day, the 11th, at daybreak, we weighed our grapnels as we got ready; we gave chase to several young black swans, most of which we caught swimming. At about 9 o'clock we observed that the boat of the hooker was missing, whereupon I changed over to another boat and sent mine down to find out where it was, and having met they came back together to me and reported that the said boat had gone ashore where they had rested themselves and had seen two blacks, a number of huts, and fires in several places.

At midday we deemed that we had reached 9 to 10 miles following the river where it was deepest, and in order to enter a bay we were forced to push the boats with our shoulders. Then we saw in the water several snakes, even though it was still a bit brackish. We noticed on the nearby embankment several footprints of people of an extraordinarily large size. We turned around about 3 to 4 miles and set foot ashore in a very attractive place. We climbed first up a little hill which could properly be given the name of 'Belvedere', since from there a very fine countryside is displayed on all sides, open land mixed with forest. We went in all directions in the hope of meeting with some of the local inhabitants, but our exertions and moves were fruitless and we were forced, through fatigue and nightfall, to go away from there in our boats, in order to find, close to the river bank, a suitable place to lie down and rest. Having found same, I also wanted to turn this place into a general rendezvous. For this reason I had sentries posted all around to guard us from surprises and dangers which could occur, and since some of the inhabitants, at the sight of our fires, might come close, I ordered the sentries to be very watchful all around.

At 2 o'clock at night, upon observing the glow of several fires, we boarded very softly and rowed noiselessly towards that place, and when we were close enough we despatched eight well-armed men with two Indians to try to catch one of the inhabitants. They went straight to these fires and when they arrived there they neither found nor saw anybody.

On the 12th from daybreak we rowed to another place where we set foot ashore and found several more huts, one foot tall on one side, and two on the other; inside them were pieces of bark from trees [for bedding], and on the outside they were covered with fine white reeds.

After progressing some distance we saw gum on nearly all of the trees, of which we took enough to serve as samples and evidence, and since I saw that our exertions would not yield further finds I resolved to go back on board, which I did without any further consultation, so that we arrived on board all together in the afternoon.

When we were there our officers told us that since our departure nothing noteworthy had happened, and that the wind had nearly always been southerly, with a topgallant breeze, from dawn to dusk.

On the morning of the 13th of January, the wind south-south-east ¼ east with a topgallant breeze, we all weighed anchor together at latitude 31°56' south and sailed in line along the coast, which reached north ¼ north-west as far as we could see, and took soundings under way, finding depths of 12 to 14 fathoms with a thick sand and sometimes pebbles.

At noon we observed that we were at latitude 31°34' south. We were rounding a reef which was calculated to be at 31°28' south, and extended into the sea a good mile along the coast. The wind was south-south-west at 6 glasses. After midday the flag of our boat was put at half-mast, because I had ordered it to go along the coast and signal to us when it saw any breaks in the reefs. At the same time we ourselves saw some surf about 1½ miles offshore. Accordingly, we allowed ourselves to run on a quarter of a mile, the water being between 10 and 14 fathoms deep, and likewise in the evening being a quarter of a mile from the said reef, we found that the depth was 18 fathoms and sometimes 20. In order not to run any risk at night along this unknown coast, we withdrew at sunset, under small sail, the wind south-south-west with a topgallant breeze. However, during the night the wind became strong enough for all sails, the depth being 30 to 45 fathoms with a sandy bottom.

At first watch the next day we sailed back towards the coast, the depth being 25, 20, and 18 fathoms, and as the cook was serving breakfast we found ourselves at the coast.

On the 14th the wind was fresh with a topgallant breeze, the depth as before with a sandy and sharp bottom and we noticed a small island or large rock which connects with a reef for a quarter of a mile along the coast towards the north.

At noon we estimated we had travelled 17 miles in 24 hours north-north-west and we observed we were at latitude 30°31' south. We then had this little island north ¼ north-west, 1½ miles away. After midday the wind was south ¼ south-east, and east-south-east with a breeze strong enough for us to set the topsails. Once again we steered out to sea, and, having cast the lead, found the depth was 20–30 fathoms, with a bottom of chalky stone with some rocks, the wind and weather being as before. We changed tack after sailing south-west for two hours and found the depth 24–21 fathoms with a chalky bottom and the wind south-south-east up to the evening. At midnight we changed tack again.

On the morning of the 15th the wind was suitable for setting the topsails and topgallants and we found ourselves close to the coast. We then changed tack, running south 2 miles offshore, where we noticed along the said coast several small islands and rocks with a long sandy reef having a strong surf.

At noon we observed that we were at latitude 30°31' [sic, should be 30°12'] south, and shortly afterwards I sent my uppersteersman with the boat and the uppersteersman of the hooker ashore because we noticed some smoke there in various places. Thus, hoping by this means to have some news of the local inhabitants, we remained until nightfall in those places until my uppersteersman and his crew came back on board with the report that they had been a good 1½ miles inland without noticing anything of importance or any person, despite walking with difficulty on the sand dunes, on which they had nevertheless observed several footprints of people and the tracks of very large birds.

In the evening we still saw fires in various places all along the coast, as far as the eye could reach without strain. The wind being northerly, we allowed ourselves to drift out to sea until the midnight watch, when we changed tack and ran in fine weather towards the land, where I sent back my steersman and 14 men to explore from dawn.

On the morning of the 16th there was a fresh northerly wind and very pleasant weather. At 5 o'clock the said boat came on board again. The uppersteersman reported that they had all gone ashore together, leaving only two men on the boat to keep watch. They had been a good three miles inland and arrived at first near a marsh where they had seen a hut nearby, and, in the vicinity, some tracks of tigers and other wild animals. They had proceeded further and approached several fires without nevertheless seeing anybody, although they had searched very thoroughly.

At 9 o'clock in the evening the wind was as before. We changed course and went out to sea until dawn next day when once again we steered towards the land.

On the 17th the wind was northerly and the weather very pleasant. Finding ourselves again near the coast we sent the boat back to the land once more, with some well-armed men.

At noon we observed that we were at latitude 30°46' south, having drifted south approximately 34 minutes in 48 hours. Because our men were ashore we held our course, coming and going between the open sea and the coast, which was similar to that in Holland, with high sand dunes. Sailing thus along the coast we passed a reef and a headland reaching out into the sea. The depth diminished from 20 to 16 fathoms with a foul bottom near the aforesaid reef or shelf, which is full of rocks and small islands.

At sunset our boat came on board, and, when the officers gave us their report, we learned that they had been about 3 hours ashore on a mountain, which appeared from the sea like Dover Castle, and observing the land from there it seemed, as far as the eye could see, to be an infertile desert. Returning, they did not find anything of importance, but it was worth mentioning that, when the boat was ¾ mile from the coast, they had visited a small island, which lay outside their route, and had found it full of young seagulls. Setting sail, we cast the lead several times and found depths of 16, 17, 18 fathoms with a foul bottom. The wind was westerly and slight, and we were steering south. At the end of the first watch the wind was northerly and the weather good. We then had a depth of 19 fathoms with the same very hazardous sea-bed. Up to the following day the wind was as before and very slight, and we found the depth was up to 27 fathoms.

On the 18th there was a fresh wind from the west. We changed course and maintained a northerly course.

At noon we observed that we were at latitude 30°51' south, about 3 miles offshore, with a depth of 23 to 24 fathoms and a very foul bottom.

Between midday and the evening the wind was west-south-west and the weather very clear and pleasant. By observation from the setting sun we found a north-west variation of 8°19'. The depth was then 26 fathoms. It was calm the whole night up to the morning.

On the morning of the 19th there was a very slight north-westerly wind. We were 3 miles offshore, and, because of the calm, we were able to approach land. At noon we changed course and came south. We then had a depth of 25 fathoms and a foul bottom.

In the evening the wind was from the south-west, and we found from a sunset observation that we had a north-west variation of 7°58', with a latitude of 30°24', being then about 2½ miles offshore. We allowed ourselves to drift all night, having braced our topsails against the wind, so as to reach land the next day. The depth was then 28 fathoms, with a very foul bottom.

On the morning of the 20th there was an easterly wind, and, finding ourselves near a flat coast, we could not go close to the land, which is why we only sent the boat to observe the reefs and dunes of the coast, and, as far as we could see, it had managed to reach land and was still without sail. Meanwhile, we passed a reef in front of which we found a depth of 24 to 23 fathoms, and, inside it a depth of 4 to 7 fathoms with a bottom sometimes stony, and, in other places, of clay with a coarse sand. Along the coast we saw several islands, linked by reefs on which the sea breaks everywhere, lying beyond the inner part of the aforementioned reef. They reach out a mile into the sea in some places, while in others they lie ½ to ¾ of a mile or only the range of a two-pounder cannon from the said reef.

All along this coast one can sail at 17, 16, and 15 fathoms. We saw our boat, which was under sail, by the coast.

At noon we observed that we were at latitude 30°20' south and the wind was mostly west-south-west. Towards evening we let ourselves drift, because our boat, which was expecting us, was coming towards us, and came aboard at sunset, the weather being really fine and the wind south-south-west. They reported to us that, having landed, they had noticed nothing of importance and that the country had appeared to them a real desert, treeless and without any other vegetation. They had found there by observation at sunset a north-west variation of 8°8'.

Then we changed course and went out to sea closest to the wind. The depth was 23 fathoms, with a foul bottom up to the midnight watch when we found a depth of 30 fathoms with the same sea-bed. We came quite suddenly to the coast, the wind coming from the south up to the morning.

On the 21st, approaching a small island, I sent the boat to reconnoitre. I found that it was surrounded by several reefs and rocks. To the north, about a mile from this island, the mainland begins.

At noon we observed that we were at latitude 29°44' south. We again sent the boat to the shore, having ordered the officers to give special attention, when setting sail, to all the sandbanks, reefs, channels, and submerged hazards in the vicinity of the coast. We found 19 fathoms of water near the vessel, but a sea-bed in no way appropriate for anchoring.

In the evening, shortly after sunset, the returning boat was not far from us and a little later came aboard. The officers reported to us that they had seen nothing of importance. We sailed out to sea again until the day watch, and then changed tack and sailed to the coast. We again sent the same boat to the shore, so that it could sail along the coast, reconnoitring it. The wind was then normal. During the day the winds from the land and from the sea still blew strongly. We had then a depth of 20 fathoms.

On the 22nd the wind was as before with a depth of 20 to 21 fathoms about 1½ miles out to sea with a very foul sea-bed. At noon we observed we were at latitude 29°5' south. We noticed a mountain to the east-north-east which was extremely high. It was the same one we had seen as being 4 miles inland, having the same appearance as Table Mountain at the Cape of Good Hope.

Before sunset our boat returned from the land and came aboard. The officers told us that there were no large reefs near the coast, as I had thought, and that they had been about a mile inland where they had found a number of patches of scrub, and that they had also noticed tracks of tigers and other wild beasts. Sailing towards the coast we found we had a depth close to the vessel of 20 fathoms, dropping to 7 fathoms. We were then close to the coast and were constrained in that place to drift out to sea until afternoon, when we changed tack and ran back once more to the coast. We had up to the following morning nearly always 22 fathoms in depth.

On the 23rd, at daybreak, I several times sent the boat ashore and to the coast to sight it. Under sail we ran close to the wind which was south along the shore at 18, 17, 16 fathoms depth on a sharp bottom.

At noon we observed that we were at latitude south of 28°10'. We continued as far as a depth of 14, 13, and 12 fathoms, and there we found the start of a better bottom, the wind as before, but the weather threatening.

In the evening our boat came back and the officers informed us that they had seen 10 blacks on shore who were quite naked and without any weapons. They were at the same latitude as at noon, 28°10', and while these blacks remained long on the shore, they made signals to our people with their feet and hands. But since there was a great surf which drew the boat along towards the shore, this stopped them from landing at once. And when they got there at last, before they had put a foot ashore, these black coastal dwellers fled at such an extraordinary speed that it was impossible for our people to follow them. And that is why they went back to their boat, with which they set sail along the shore for about a mile where they once again went ashore. And having advanced about ½ mile on land, they had come to a certain water, salty and red like blood, and at the edge of this water there was salt in abundance. Superficially it looks as though at the time of full or new moon or in the months of March and September the sea rises higher, causing the salty water to throw itself into these places and that afterwards through the nature of the bottom, which is red, and through the power of the sun, the salt is produced according to the quantity of the water. Besides this they had found nothing noteworthy, and while returning on board they had sounded from 4 to 12 fathoms depth.

By nightfall we observed a very steep coastline about 2 miles inland and we arrived in a bay where we dropped anchor each at a cannon-shot distance from the others and close to shore with a good bottom, depth of 19 fathoms, the wind as before.

All night the wind was the same, with very fine weather up to the following morning. On the morning of the 24th the wind was southerly with a topsail breeze. The skippers of the other vessels were requested by signal to come on board my ship.

I sent the boat to make soundings in this bay and also to reconnoitre an appropriate place for a safe landing. The officers of the other vessels came aboard and decided it was opportune to send some armed men ashore to reconnoitre the countryside, which appears so pleasing in this place, through its hills, valleys, and plains. Thus, in order not to waste the opportunity, it was decided to prepare from this evening all necessary provisions. This is why we ordered 36 well-armed men with some officers to go ashore to proceed into the country shortly before dawn, that they take provisions for eight days, and that 14 sailors should also hold themselves ready to go into another place with the boat.

The wind and weather as before until morning. At daybreak my uppersteersman and the bookkeeper with the people selected had left ship and rowed strongly straight to the shore, having been advised by the men sent out the previous day that the bay was everywhere safe.

On the 25th the said uppersteersman and bookkeeper came back on board unexpectedly and reported to us that they had not seen anything except some huts and several tracks of men and beasts, and that, near a marsh, they had noticed an area 10 to

12 feet across full of frogs. The wind was south-south-west the whole night and south-south-east in the morning.

On the 26th I again sent some people ashore. They returned in the evening reporting that they had seen nothing except some bush and patches of scrub. The wind was south-south-east and south-south-west up to the morning.

On the 27th there was a southerly wind and fine weather. After breakfast the people who had been ashore came aboard again and told us that they had been 5 miles inland without having seen anything of importance or come across anything but huts, and that, because they had not found any water, they had been forced to dig holes in the ground, in which the water seemed to them to be a bit salty, but thirst had compelled them to drink it. They also reported that this country was very difficult to cross because of thick and thorny scrub. In the evening, the night, and up to the following morning, the wind was as before.

On the 28th the wind was southerly with a topsail breeze. After breakfast we raised anchor and ran under sail straight out to sea, because, through the height of the land, we could not take the noon elevation of the sun there.

At noon we observed that we were at latitude 27°40' south with a depth of 30 fathoms at half a mile from land. Sailing along the coast we noticed that the land before us was fringed by rocks and reefs, against which the waves broke strongly, so that it was almost impossible to land. Thus, sailing on, we ran from a depth of 22 fathoms up to 24 fathoms with a safe bottom at 1½ miles from the shore.

In the evening at sunset we went under small sail, having a depth of 40 fathoms, a very good bottom, and a topsail breeze as before. It remained the same all night.

In the morning we ran back to the coast and had a depth of 58 fathoms with a good sandy bottom. One has this depth or less nearly all along this coast of the South Land. Thus, by taking good care to take soundings, one need fear no danger.

On the morning of the 29th, the wind was south-south-east, the weather heavy, with a topgallant breeze. We had a depth of 50 to 40 fathoms, and going under sail, sometimes 36 fathoms, being then half a mile offshore. The wind then veered south-west.

The land here appears very bleak, and so abrupt as if the coast had been chopped off with an axe, which makes it almost impossible to land. The waves break here with so great a fury that one should say that everything around must shake and become dismembered, which appears to us a truly terrible sight. Nevertheless, we found it rather difficult to drag ourselves away from this evil place.

We observed that we were at latitude 26°36' south, the wind as before up to the evening. We ran with a depth of 25–26 fathoms along the coast 1½ miles offshore. We could maintain a westerly course, and, because of that, we sailed at sunset into the open sea with a southerly wind. Towards the north-west the coast reaches out a little further west. At 10 o'clock in the evening our comrades were either down-wind or up-wind from us and we reduced sail a little and let ourselves drift for a while.

At three glasses of the day watch, we ran along the coast under small sail up to daybreak. We could not see our other two vessels. There was a coastal wind from the south with a topgallant breeze. The land reaches out north-west here and is extremely steep all along the coast. We were unable to see any gap, but were very much aware of the huge surf. The depths here are 35, 36, and 40 fathoms up to the third glass of the watch before noon.

On the 30th we came to the Roads of Dirk Hartog, where we thought we would be able to anchor with a safe bottom at 16, 15, and 14 fathoms. However, we found, on the contrary, that it was a foul and stony bottom. This is why we took advantage of a following wind.

At noon, we observed that we were at latitude 26°12' south. At about 2 o'clock we reached a place with a depth of 18 fathoms with a very good sandy bottom, where we anchored. We still had not seen again either the galliot or the hooker. We then identified a reef lying across the bay, and to take good soundings and reconnoitre it, we sent our uppersteersman and understeersman there with the two boats with instructions to make a thorough examination. After a short period they came on board again and told us that everywhere over the said reef there was three fathoms of water, but that the bottom was extremely foul and dangerous. Towards evening the hooker and the galliot arrived from the south coast and anchored near us. We had the southern headland of the said bay to the south and south ¼ south-west at about a cannon's range from us.

At night the wind was southerly with a topsail breeze and this continued until the morning of the 31st, the last day of January.

A little later the wind came from the south ¼ south-west with the same strength as before, but the sea was very high with huge waves.

I sent my bookkeeper with my uppersteersman and the steersman of the hooker with the boat to go towards the land to reconnoitre a river that I presumed to be there. They came on board again in the evening and reported that there was an island in this place which was estimated to reach out to sea north-north-west, and a gulf which reached south and which was very deep everywhere [*Geelvinck* log says very shallow]. This was the reason why they had not been able to use the boat to set foot ashore. The wind remained as before throughout the night up to daybreak.

On the first day of February the wind was as before with very fine weather. For the second time I sent the boat with the aforesaid steersmen to the said island with orders to examine whether one could approach with the vessels and anchor there. At noon up to the evening the wind was as before. All night it was the same and this continued on the 2nd and 3rd the wind veering towards the south-east in the mornings with very changeable weather. At noon on the following day the wind was southerly with the same strength as before. At about two o'clock in the afternoon the boat returned with the people I had sent out on the 1st of February to reconnoitre the said island. My uppersteersman reported to me that it stretched a good 2 miles east and 2 miles north-east ¼ north, and that, moreover, it stretched 11 or 12 miles along the coast. Sailing along this coast they had 52 fathoms to start with, but, after making further progress, they had found very uneven depths of 2½, 3, 4, 5, 6, and 7 fathoms. The sea-bed had sometimes been sandy, and at others, rocky. They had estimated the width of the bay to be about 6 to 7 miles. The day being over, they had spent 10 hours at night anchored behind a low headland, where they had found a quantity of turtles, of which they brought two on board.

On the 2nd in the morning the wind south, topsail breeze, fine weather. They had weighed the grapnel and sailed ahead until two hours after sunrise. Having arrived near a headland of the coast, they put the boats on the shore of the island and stepped ashore where they had first seen a high mountain, which they climbed in order to obtain a much wider view of the land and the surroundings where there might be some debris of ships. But they saw nothing noteworthy excepting a post fixed on the said mountain. They went over to it and found a pewter dish on which had been written the year 1616 the 15th [sic, actually the 25th] of October the vessel the Eendracht of Amsterdam on which was the upper-merchant Gillis Miebais of Liege and commanded by skipper Dirk Hartog of Amsterdam with the under merchant Jan Stins and the first officer Pieter Dookes has been here and set sail again the 27th ditto for Batavia. From this I learnt that I was truly in the Bay of Dirk Hartog, of which I have often heard tell.

The wind was south-south-east and south-south-west all night up to the morning of the 4th. The wind was then south-south-east with the same strength as before. We weighed anchor and left this roadstead. Setting sail, we went to Dirk Hartog's Bay. The wind was southerly. The coast stretches north-west, the depth 38 fathoms with a stony bottom, being then half a mile offshore. At noon we observed that we were at latitude 25°27' south [the *Geelvinck* log says 25°37']. The coast stretches north-north-west and we anchored at 4 o'clock in the afternoon in the said roadstead with a depth of 17 fathoms and a very fine sandy bottom.

We observed the southern headland to be west ¼ south-west about a quarter of a mile from us, stretching out from the coast about a cannon's range.

The north coast was 5 miles away from us. This roadstead is the cleanest and best which may be encountered in Terra Australis, because it has an even sea-bed right up close to the shore. The depth here is 12 to 10 fathoms. One sees here successive vistas of sand dunes and bare, treeless mountains. People who like catching turtles can well satisfy their liking here. The wind was southerly all night with fine weather.

On the 5th the wind was south ¼ south-east with a topsail breeze. I decided to sail to shore with my boat and that of the hooker, taking with me skipper Colaert with his understeersman and the third mate of my vessel. We took provisions for eight days. At 3 o'clock in the afternoon we left ship, the wind south ¼ south-west with a topsail breeze. In our boat we had 14 well-armed men. After sailing for a long time in one direction without yet sighting land, we sailed in the other direction until one glass after sunset when we reached the said island. Estimating that we had sailed for 2 miles into the wind, we stepped ashore to spend the night there. The wind was south ¼ south-east, a little more southerly than before, with a topsail breeze, and this continued all night.

On the morning of the 6th, the wind was southerly with a topsail breeze. We weighed the grapnel, set sail, and ran east-south-east towards the coast. The sea began to rise and it seemed that there were reefs in some places.

At 10 o'clock in the morning we could still not see the mainland. We changed tack once again, sailing towards the said island, where we arrived a little after midday.

After we had all eaten, we fastened the two boats together and thus made them pull along the coast [presumably this means that they rowed them]. In the water we saw an animal in the shape of a dog [perhaps a small dugong] and several turtles. We spent the whole day in this way without making progress because of the stiff breeze and the sea was so rough that we did not dare to attempt a landing. However, in the evening we went ashore in order to sleep. We estimated that we were about 6 miles from our vessel. During the whole night the wind and weather were as before.

A little after sunrise on the 7th we all embarked in our boats. After we had passed a very large reef, the wind began to blow strongly, and we had as much of it as we could bear. The boat of the hooker changed tack to sail with a following wind, and for that reason, it drew away from us up to midday, when it came to rejoin us. Meanwhile, we had caught a ray or a similar type of fish. After midday we had a stiff topsail breeze. In the evening we reached the outer end of the island to spend the night there. All night the wind and weather were the same.

On the morning of the 8th, the wind being southerly, we again set sail. On the stroke of 9 o'clock, we spotted the mainland at south-east ¼ south and set our course south-east and east-south-east.

After midday we came close to a river. We proceeded about 3 miles further south with a strengthening wind, accompanied by loud thunder and frequent flashes of lightning. This is why we sailed with a following wind towards the northern headland, where we went ashore to sleep. All night long the wind was easterly, and there were flashes of lightning.

On the 9th, the wind was from the south-east. We set sail again and ran along the coast towards a river which extended inland east ¼ east. After going up river for 1 mile, we found that it petered out there. We went ashore to sleep. The wind was southerly with a topsail breeze. This continued the whole night, with heavy rain, thunder, and lightning.

On the morning of the 10th the wind was from the north-east, veering further north, with a full sail breeze. After sailing for about 3 miles along the coast, we steered again towards the island.

At noon the wind was very variable. At 9 o'clock in the evening we approached a low headland, where there was a strong surf. The weather began to improve, and we disembarked in order to sleep.

On the 11th the wind was northerly, slight, and calm. At 7 o'clock in the morning, I came back on board with my boat. The senior officers from each vessel came on board to report to me what had occurred during my absence. On Wednesday the 6th the wind was southerly, and they had not been able to take observations because they were too close to the land. At night the wind was south ¼ south-east with a breeze suitable to carry the lower sails and the topsails. They had found that the current was running strongly towards the coast.

On the 7th the wind had been south ¼ south-east with a topsail breeze. After midday they had held council and decided to send some men ashore. My bookkeeper and my understeersman went there with the boat of the hooker and that of the galliot in order to reconnoitre the said island as much as possible. But, unfortunately, a gust of wind from the land had overturned the last boat, and afraid that the same accident might befall the other boat, the people in it had turned around quickly to obtain help for their comrades. However, the misfortune had been noticed from the other vessels, which had immediately fired three cannon and put the flag at half mast in order to inform all the boats that were ashore of the calamity which had just occurred. These men, having noticed this signal, rowed there with all their strength. However, the galliot, having weighed anchor and set sail, arrived before them and saved all the men, except for a carpenter, who had drowned. The others had been saved by a line which had been let out from the galliot.

From midday to the evening and all night the wind was south-south-west with a topsail breeze.

On the 8th the wind had been variable and calm. All morning my uppersteersman had been at a point of the island about a full mile from the headland which juts out into the sea. He tried to reconnoitre the extent of a neighbouring reef and found it extended for ¾ mile from the headland across the bay, having to the north of the said reef beyond the bay 2, 2½, and 3 fathoms in depth. Really big vessels can easily enter [the bay] with depths of 10, 12, 14, and 15 fathoms, the bottom being reasonably good.

From midday to the evening the wind was from the south-west with a topsail breeze. At night the wind was south ¼ south-east and south-south-east. They had sent the boat ashore to catch some turtles there and saw that there was a strong surf along the coast.

On the 9th the wind had been from the south-east, the breeze scarcely strong enough to set the topgallants. Nevertheless, the sea coming from the north was getting much rougher, and its motion alone, causing strong surf at the coast, was capable of shaking the vessels. That is why they all decided to save themselves from harm and depart from this coast. They hurried so far out to sea towards the north that they had the headland of the said island west ¼ south-west and west-south-west about 2 miles away from them. They dropped anchor in a depth of 17 fathoms on a really good sandy bottom.

In the afternoon the wind was south ¼ south-east. They sent the boat with an anchor and some provisions to the shore in the direction of the hooker's boat, which was at the eastern point of the island. When they both returned at night they learned from them that there had been a great deal of loud thunder out at sea. The wind was south-south-east and south with a topgallant breeze and a high sea running from the north up to the following morning.

On the 10th there was a slight breeze from the north, and before midday the breeze was from the north-east and also quite slight. At noon they had observed the latitude was 25°16' south. In the afternoon the wind had been variable and calm and the same all night right up to the following morning when this report was made to us on our return from the land.

I then had a pewter plate flattened on which had been scratched on one side as already has been told A° 1616 on 25th October arrived here the ship de Eendracht from Amsterdam etc.

On the other side [sic, actually on the same side] I had the following written: A° 1697 the 4th of February arrived here the ship Geelvinck from Amsterdam on which commanded Skipper Willem de Vlamingh of Vlielandt, the assistant Joannes Bremer of

Copenhagen, and the uppersteersman Michiel Bloem of Bremen, the hooker the Nyptangh on which commanded skipper Gerrit Colaert of Amsterdam and the uppersteersman Gerrit Gerritsen of Bremen and the galliot 't Weseltje commanded by Cornelis de Vlamingh of Vlielandt and the steersman Coert Gerritsen of Bremen, having departed again on the 12th for more extensive exploration of the Southland and bound for Batavia.

On 12th February 1697 the wind was variable. Nevertheless, shortly before noon, with the wind west-south-west, we all decided to weigh anchor, and after midday, we set sail, the wind being south-south-west with a topsail breeze. We ran freely for a good mile along the coast.

In the evening at sunset we observed that the small bay or inlet, where we had anchored, was to the west [*Geelvinck* log says east] of us. We then found the reef, which was just as it had been described on the 8th of this month, and on which there was a strong surf. The wind then veered to the south-west with a very fine breeze. We were 1½ miles offshore and had a depth of 40 fathoms. All night, up to the following morning, we ran south close to the wind.

Early in the morning of the 13th, the wind was south-south-west with a topsail breeze. We had depths of 43, 45, and 47 fathoms with a sandy bottom. We changed tack for a following wind, and then sailed east-south-east towards the coast.

At breakfast we were 3 miles offshore. Before midday the wind was south-west with a topsail breeze and we had depths of 35, 33, 21, and 19 fathoms. We were then past the headland that had been to the east of us, and which we had seen previously. We ran into a little bay that we had met on setting our course. We found depths of 15 and 14 fathoms and sometimes of only 13 fathoms at about a mile offshore, and, at noon, we were at latitude 24°38' south. After midday the wind was south-west and strong enough to carry the topsails. We had depths of 15 to 12 fathoms with a very fine bottom, and we all anchored.

One hour before sunset we were at anchor in 12 fathoms of water and could still not make out the land, which we estimated to be about eight miles away. All night up to the following morning the wind was southerly with a topsail breeze.

On the morning of the 14th we weighed anchor and set sail. Steering to the south-east ¼ east, we found the depth diminishing from 12 to 5½ fathoms. After sailing for 4 glasses, we sighted land to the east-south-east about 2 miles away from us. We had a depth of 4½ fathoms with a very good bottom. After running east-south-east for 2 miles, we changed tack and ran west. We could maintain a course of west ¼ south-west, finding a depth of 8 fathoms increasing to 15 fathoms. Before midday the wind was southerly, with a topsail breeze. We estimated that we had run west ½-point further south for 3½ miles. At noon we observed that we were at latitude 24°51' south. We then saw the south coast west-north-west about 3½ miles away.

After midday we noticed the inlet of the place we had left or passed and where there was a strong surf, its arms reaching out into the sea. We found the depth diminishing from 14 fathoms to 7 fathoms and afterwards increasing back to 14 fathoms. We changed course, sailing south-east with the depth increasing from 9 to 12 fathoms and everywhere there was a good sea bed. At sunset we estimated that we had run south-east for about 3½ miles. We then had a depth of 13 fathoms with a bottom of sand and shells. We anchored in this place, but still could not see the land. The wind was southerly with a topsail breeze, and continued thus until nightfall. At sunset we set sail with a very fine topsail breeze.

The wind was as before and we had a depth diminishing from 10 to 6 fathoms. The hooker was still anchored, because it had lost a fluke from its anchor. After we had sailed on for a mile, we were obliged to change course in order to see what was troubling it. When we had come close to it we found it was trying to raise the other anchor, which had become badly snagged. Thus we spent the time coming and going until the following morning.

On the morning of the 15th we took down our lower sails. With a southerly wind and topsail breeze, we found, while sailing, depths of 7 to 13 fathoms and upwards of 14 fathoms with a good sandy bottom. At noon, after we had run 3 miles, we noticed at the coast a hill lying west ¼ north-west from us about 3 miles away and situated at latitude 24°36' south [*Geelvinck* log says 24°56'].

After midday we still held a westerly course close to the wind, which was south ¼ south-west with a topsail breeze. We had depths of upwards of 5 fathoms, of 10 and 12 to 13 fathoms as formerly, up to 1 glass before sunset, when we anchored. We calculated that we had sailed 4 miles south-east ¼ east, the sea-bed as above, as also the wind and weather.

The night as before up to the following morning, when I gave the order to weigh anchor. We set sail very quickly. We sailed east and could maintain a course of east ¼ south-east. After we had sailed for 2 glasses and moved forward 1 mile towards the east ¼ north-east, we changed tack. We had depths of 8 to 9 fathoms. In the evening the wind was south ¼ south-east. We reefed our topsails and sailed thus all night up to the following morning.

On the 16th February the wind veered southerly with a very fine topsail breeze. Before midday we had a depth of 9 to 10 fathoms. We held our course south ½-point further east and were about 4 miles offshore. After midday we noticed a point to the east-south-east 3 miles away from us. At noon we had observed that we were at latitude

25°6' south. At about four hours after this observation we arrived close to a headland stretching out into the sea which was of the said bay of Dirk Hartog about south and north, the one near the other 6 miles.

At the same time we noticed a reef to the south-west ¼ south of the said headland on which there was a strong surf. However, one could quite easily pass between them without danger. After arriving in the open sea, we steered north along the coast. The wind was southerly during the night. After running four glasses of the first night watch, we sailed out to sea and let ourselves drift up to one hour into the day watch, when we changed tack and sailed east north-east towards the coast. There was a topsail breeze and we had depths of 48, 30, and 25 fathoms up to quite close to the shore.

As the sun began to rise on the 17th the strength of the breeze was as before.

At noon we sailed into water 20 fathoms deep quite close to the coast. We observed that we were at latitude 23°50' south. The land is very high here and nearly everywhere precipitous. The coast stretches north ¼ north-east. In the afternoon we passed a headland and the land in the region is low, extending north-north-east. The wind was as before with a topsail breeze. We ran along the coast in 20 and 21 fathoms of water. The coast, as it stretches away, gets higher and higher for about 10 miles.

At sunset we ran out to sea close to the wind. We had a depth of 23 fathoms with a good sandy bottom. At night we noticed many gleams of fire at various places on the mainland, which even continued up to daybreak. We sailed east towards the coast, and came abruptly very close to a reef which stretches out to sea from the mainland north and north ¼ north-east of what we estimated to be latitude 23° south. We ran along the coast up to daybreak with depths of 22, 23, and 24 fathoms.

On the 18th the wind was southerly and weather as before. At midday we noticed some reefs to the north-west ¼ west of us. Shortly before midday, the coast reaching north-west and getting increasingly steep, we noticed, at latitude 22°31' south, a rock or reef jutting out from the water.

The coast is rather high here, but nevertheless it is quite easy to go ashore at any point along the land, which stretches north-east ¼ north. There is a small reef about half a mile from this coast. We had depths of 38 to 40 fathoms.

At sunset we ran out to sea for up to 5 glasses of the midnight watch. On the morning of the 19th, after I had given the signal, we all changed tack to go towards the land. At breakfast we were about ¾ mile away from it. When we were ½ mile from the shore we noticed a reef. We had the same wind as before.

At noon we came across a cove, and after making a survey, we found a very fine roadstead, and, shortly afterwards, anchored there in 10 fathoms, the bottom being of fine sand. Meanwhile, the hooker anchored a good mile to the north of us near another point. The skipper of the galliot came to report to me that he had observed that he was at latitude 21°32' south. In the evening and all night we had very fine weather. We saw, as usual, several fires and plumes of smoke on the mainland. At 3 glasses of the day watch on the 20th, I gave the signal and we all weighed anchor together and set sail. We set our course towards a cove situated behind the point where the hooker had been anchored. We now saw this vessel coming in our direction. However, we continued sailing east ¼ north until we could barely see it. We then steered south-east and south-south-east in order to reconnoitre the said cove as much as possible.

At noon we observed that we were at latitude 21°39' south. We noticed three small islands to the south-east spreading out 5 miles offshore. Near there we had a depth diminishing from 10 to 5 fathoms with, sometimes, a good bottom, and, at other times, a bad bottom. At 4 glasses after midday we changed course towards the north and could maintain a course north-north-west. We then noticed the hooker which was sailing among some small islands which spread out in this place towards the south-east. While waiting for us it was drifting, and, after we had joined it, we both anchored. It was shortly before sunset with a depth of 5 fathoms and a good sandy bottom. The wind was southerly and remained the same until the following day.

Early in the morning of Thursday 21st February 1697, the wind was as before with a topsail breeze but we could not set sail because of certain winds from the sea which prevented us from leaving until midday.

I then gave the signal to weigh anchor and to leave this South Land for good, because the monsoons were over and I concluded that I had done enough in the interests of the company and its directors, my lords and masters, even hoping that they would continue to employ me, and that I could still convince them that I am a very assiduous and devoted servant. I also concluded that it was useless to spend more time exploring and reconnoitring the Terra Australis, believing that I had obtained sufficient information to pass on to them to enable them to send another expedition to this coast, without holding any fear for their vessels, which is a rather important indication of the success of my voyage. Finally, I believed that I had sufficiently fulfilled my duty to my employers, and even to myself, since I had followed and respected the orders and instructions I had been given, such as the making of charts of all these coasts and examining them, as well as my logs and manuscripts, from which it could be seen that, regardless of perils and dangers, I had done everything possible to explore this Terra Australis and its surroundings.

Because the season was advanced and my intention was to reach Batavia as soon as possible, we set full sail and steered towards an island in order to place there our fixed point, our position then being latitude 21° south and longitude 131°22'.

Towards evening the wind was from the south-west with a topsail breeze. We steered north-west. The wind was the same all night up to the following morning. We could no longer see the South Land, and we corrected our compass 25° north-west.

Bibliography

Anon. 1701, *Journaal Wegens een Voyagie Gedaan op Order der Hollandiche Oost-indische Maatschappy in de Jaaren 1696 en 1697 door het Hoeker-scheepje de* Nyptang, *het schip de* Geelvink, *en het galjoot de* Wezel, *na het Onbekende Zuid-Land en Wyders na Batavia*, Willem de Coup, Willem Lamsvelt, Philip Verbeek & Jan Lamsvelt, Amsterdam. (Note: This publication was derived from the log of the *Nyptangh*, written by Mandrop Torst).

Appleyard, R.T. & Manford, T. 1979, *The Beginning: European discovery and early settlement of Swan River, Western Australia*, University of Western Australia Press, Nedlands.

Balbian Verster, J.F.L. de 1899, ''t Onbekende Zuytlandt (de geschiedenis van twee tinnen schotels)', *Eigen Haard*, no. 35, pp. 556–60.

Bassett, M. 1962, *Realms and Islands*, Oxford University Press, London.

Bates, Daisy 1992, *Aboriginal Perth: Bibbulmun biographies and legends*, ed. P.J. Bridge, Hesperian Press, Carlisle, WA.

Battye, J.S. 1924, *Western Australia: A history from its discovery to the inauguration of the Commonwealth*, Clarendon Press, Oxford.

Blainey, G. 1995, *A shorter history of Australia*, Reed Books, Port Melbourne.

Crowley, F.K. 1960, *Australia's Western Third: A history of Western Australia*, Heinemann, Melbourne.

Crowley, F.K. & de Garis, B.K. 1969, *A Short History of Western Australia*, Macmillan, Sydney.

Dana, R.H. 1840, *Two Years before the Mast*, ch 35, Harper, New York.

Edwards, H. 1962, *Gods and Little Fishes*, Peter Davies, London.

Flinders, M. 1814, *A Voyage to Terra Australis, in the Years 1801, 1802, and 1803*, G. & W. Nicol, London.

Freycinet, L.C.D. de 1825, *Voyage autour du Monde, Exécuté sur les Corvettes l'Uranie et la Physicienne pendant les Années 1817–1820*, Imprimerie Royale, Paris.

Green, N. 1984, *Broken Spears*, Focus Education Services, Perth.

Grey, G. 1841, *Journals of Two Expeditions of Discovery in North-West and Western Australia During the Years 1837, 38, and 39*, 2 vols, T. & W. Boone, London.

Halls, C. 1965, 'The loss of the *Ridderschap van Holland*', *The Annual Dog Watch*, no. 22, pp. 3–8.

Hammond, J.E. 1933, *Winjan's People*, ed. Paul Hasluck, Imperial Printing, Perth.

Heeres, J.E. 1899, *The Part Borne by the Dutch in the Discovery of Australia 1606–1765*, Luzac & Co., London, & E.J. Brill, Leiden.

Henderson, G. 1980, *Unfinished Voyages: Western Australian shipwrecks 1622–1850*, University of Western Australia Press, Nedlands.

Henderson, J.A. 1982, *Marooned*, St George Books, Perth.

Hordern, M.C. 1997, *King of the Australian Coast: The work of Phillip Parker King in the* Mermaid *and* Bathurst*, 1817–1822*, Melbourne University Press, Carlton.

Horner, F. 1987, *The French Reconnaissance: Baudin in Australia, 1801–1803*, Melbourne University Press, Carlton.

Jacobs, E.M. 1991, *In Pursuit of Pepper and Tea: The story of the Dutch East India Company*, Netherlands Maritime Museum, Walburg Pers, Zutphen, the Netherlands.

King, P.P. 1827, *Narrative of a Survey of the Intertropical and Western Coasts of Australia, Performed in the Years 1818 and 1822*, John Murray, London.

Leupe, P.A. 1868, *De Reizen der Nederlanders naar het Zuidland of Nieuw-Holland in de 17e en 18e Eeuw*, Amsterdam.

Lines, W.J. 1994, *An All Consuming Passion*, Allen & Unwin, Sydney.

Loos, N.A. 1994, 'Aboriginal–Dutch relations in north Queensland, 1606–1756', *Queensland Heritage*, vol. 3, no. 3, pp. 3–8.

Major, R.H. 1859, *Early Voyages to Terra Australis, Now Called Australia*, Hakluyt Society, London.

Marchant, L.R. 1982, *France Australe*, Artlook Books, Perth.

Marchant, L.R. 1988, *An Island unto Itself: William Dampier and New Holland*, Hesperian Press, Carlisle, WA.

Moore, G.F. 1884, *Diary of Ten Years of an Early Settler in Western Australia*, Walbrook, London.

Playford, P.E. 1996, *Carpet of Silver: The wreck of the* Zuytdorp, University of Western Australia Press, Nedlands.

Playford, P.E. 1998, 'Willem de Vlamingh's exploratory voyage to the Southland in 1696–97', in *The ANCODS Colloquium*, eds Jeremy Green, Myra Stanbury and Femme Gaastra, Special Publication no. 3, Australian National Centre for Excellence in Maritime Archaeology, Fremantle, p. 19-27.

Prévost d'Exiles, A.F. 1758, *Histoire générale des voyages…*, vol. 15 (of 25 vols), Avec approbation et privilège du Roi, Paris.

Renié, F. 1944, 'Les inscriptions de la Baie des Chiens-Marins (Nouvelle-Hollande Occidentale)', Comptes rendus de l'Académie des Inscriptions et Belles Lettres de l'Institut de France, Séance du 22 décembre 1944, pp. 505–15.

Robert, W.C.H. 1972, *The Explorations, 1696–1697, of Australia by Willem de Vlamingh*, Philo Press, Amsterdam.

Schilder, G. 1976a, *De Ontdekkingsreis van Willem Hesselsz. de Vlamingh in de Jaren 1696–97*, Martinus Nijhoff, The Hague.

Schilder, G. 1976b, *Australia Unveiled: The share of the Dutch navigators in the discovery of Australia*, Theatrum Orbis Terrarum, Amsterdam.

Schilder, G. 1978, 'A continent found by mistake', *The Geographical Magazine*, February 1978.

Schilder, G. 1984, *The Southland Explored: The voyage of Willem Hesselsz. de Vlamingh in 1696–97*, Uitgeverij Canaletto, Alphen aan den Rijn, Netherlands.

Schilder, G. 1985, *Voyage to the Great South Land: Willem de Vlamingh, 1696–1697*, Royal Australian Historical Society, Sydney.

Scott, E. 1916, *A Short History of Australia*, Oxford University Press, London.

Seddon, G. 1970, *Swan River Landscapes*, University of Western Australia Press, Nedlands.

Seddon, G. 1972, *Sense of Place*, University of Western Australia Press, Nedlands.

Seddon, G. 1995, *Swan Song: Reflections on Perth and Western Australia, 1956–1995*, Centre for Studies in Australian Literature, University of Western Australia, Nedlands.

Sigmond, J.P. & Zuiderbaan, L.H. 1976a, *Nederlanders Ontdekken Australië*, Unieboek BV, Bussum, Netherlands.

Sigmond, J.P. & Zuiderbaan, L.H. 1976b, *Dutch Discoveries of Australia*, Rigby, Adelaide.

Somerville, W. 1948, *Rottnest Island, its History and Legends*, Rottnest Island Board, Perth.

Stapel, F.W. 1937, *De Oostindische Compagnie en Australie*, van Kampen & Zoon, Amsterdam.

Valentyn, F. 1726, *Oud en Nieuw Oost-Indien*, Derde Deel, Tweede Stuk, Dordrecht–Amsterdam, pp. 68–71.

Witsen, N. 1705, *Noord en Oost Tartarye*, Tweede Druk, vol. 1, François Halma, Amsterdam, pp. 179–83.

INDEX

The Author

Phillip Playford has received national and international recognition as one of Australia's leading geologists, and is also well known as a historian. In 1998 he won a Western Australian Premier's Book Award for his book '*Carpet of Silver, the wreck of the Zuytdorp*', and was appointed as a Member of the Order of Australia, for service to geology and to historical research on early Dutch exploration and shipwrecks in Australia.

Dr Playford's interest in Willem de Vlamingh first resulted from his research on the *Zuytdorp*, and his visit, in 1955, to the place on Dirk Hartog Island where Vlamingh found Dirk Hartog's famous plate. He decided to write this book as a result of the Willem de Vlamingh Tricentennial commemoration, held in 1996-97.

Terra Australis:

'a paradise on earth' and 'an infertile desert'

Until now, knowledge of Willem de Vlamingh's expedition to Terra Australis in 1696-97 for the Dutch India Company has been based on the ships' logs. Vlamingh's personal journal was lost after it was translated into French during the early eighteenth century. Phillip Playford's recent rediscovery of this translation, preserved in a Paris archive, has shed new light on Vlamingh and his voyage of discovery.

Willem de Vlamingh's observations show that he was in two minds about Terra Australis. On the one hand, he was enthralled by Rottnest Island, describing it as a 'paradise on earth', and by the beauty, fine climate, and attractive trees of the Swan River area. On the other, he was unimpressed by the coastal region to the north, reporting that it consisted of 'infertile desert'. His discoveries in the Swan River area stimulated interest by the French, and later the British, leading eventually to British colonisation at Perth in 1829.

Willem de Vlamingh